HAM
RADIO Made Easy!

Edited by:
Steve Ford, WB8IMY

Published by: The American Radio Relay
League • 225 Main Street
Newington, CT 06111-1494

Foreword

As a newly licensed ham, the last thing you want to do is wade through page after page of tedious indoctrination. I hardly blame you. No doubt you did plenty of reading while preparing for your exam. Now you have your call sign and you want to get on the air . . . *immediately*!

But passing your license examination is only the *beginning* of the learning process. You have a whole universe of possibilities open before you. New questions are coming in waves . . .

- Which transceiver should I buy? What about an antenna?
- How do I put everything together and make it work?
- How do I contact someone across the state, the country, the *world*?
- When I pick up the microphone, CW key, keyboard, or video camera (yes, video camera!), what am I supposed to do?

You have two choices: (1) You can stumble through this vast hobby and learn from many painful (and expensive) mistakes, or (2) Do some additional reading and prepare yourself.

Obviously, we favor the latter choice!

In this book Steve Ford, WB8IMY, takes you on a breezy journey through Amateur Radio and arms you with the tools you need to get the most out of the hobby. Unlike many books of this type, Steve mixes serious instruction with a generous dose of humor and opinion. You'll read it to get started, then keep it handy as a valuable reference. This is the kind of how-to book that never goes out of style.

Welcome to Amateur Radio!

David Sumner, K1ZZ
Executive Vice President

Newington, Connecticut
November 1995

About The American Radio Relay League

The seed for Amateur Radio was planted in the 1890s, when Guglielmo Marconi began his experiments in wireless telegraphy. Soon he was joined by dozens, then hundreds, of others who were enthusiastic about sending and receiving messages through the air—some with a commercial interest, but others solely out of a love for this new communications medium. The United States government began licensing Amateur Radio operators in 1912.

By 1914, there were thousands of Amateur Radio operators—hams—in the United States. Hiram Percy Maxim, a leading Hartford, Connecticut, inventor and industrialist saw the need for an organization to band together this fledgling group of radio experimenters. In May 1914 he founded the American Radio Relay League (ARRL) to meet that need.

Today ARRL, with more than 170,000 members, is the largest organization of radio amateurs in the United States. The League is a not-for-profit organization that:
- promotes interest in Amateur Radio communications and experimentation
- represents US radio amateurs in legislative matters, and
- maintains fraternalism and a high standard of conduct among Amateur Radio operators.

At League Headquarters in the Hartford suburb of Newington, the staff helps serve the needs of members. ARRL is also International Secretariat for the International Amateur Radio Union, which is made up of similar societies in more than 100 countries around the world.

ARRL publishes the monthly journal *QST*, as well as newsletters and many publications covering all aspects of Amateur Radio. Its headquarters station, W1AW, transmits bulletins of interest to radio amateurs and Morse code practice sessions. The League also coordinates an extensive field organization, which includes volunteers who provide technical information for radio amateurs and public-service activities. ARRL also represents US amateurs with the Federal Communications Commission and other government agencies in the US and abroad.

Membership in ARRL means much more than receiving *QST* each month. In addition to the services already described, ARRL offers membership services on a personal level, such as the ARRL Volunteer Examiner Coordinator Program and a QSL bureau.

Full ARRL membership (available only to licensed radio amateurs) gives you a voice in how the affairs of the organization are governed. League policy is set by a Board of Directors (one from each of 15 Divisions). Each year, half of the ARRL Board of Directors stands for election by the full members they represent. The day-to-day operation of ARRL HQ is managed by an Executive Vice President and a Chief Financial Officer.

No matter what aspect of Amateur Radio attracts you, ARRL membership is relevant and important. There would be no Amateur Radio as we know it today were it not for the ARRL. We would be happy to welcome you as a member! (An Amateur Radio license is not required for Associate Membership.) For more information about ARRL and answers to any questions you may have about Amateur Radio, write or call:

ARRL Educational Activities Dept
225 Main Street
Newington CT 06111-1494
(860) 594-0200
Prospective new amateurs call:
800-32-NEW HAM (800-326-3942)

Contents

CHAPTER 5: OUTTA SPACE!

Lots of hams enjoy satellite communication, even though this segment of our hobby lacks "atmosphere."

CHAPTER 6: THE CAMERA NEVER LIES

Expose your true self to the world through your own amateur television station!

CHAPTER 7: WORKING THE WORLD ON A WIRE

The public calls it "shortwave." Hams call it "HF." It's a world where a couple of watts of power and a piece of wire can take you anywhere.

INFO GUIDE

Like an overstuffed sandwich, we have pages and pages of useful information.

INDEX

1

The Spirit of Radio

Invisible airwaves crackle with life,
Bright antenna bristle with the energy.
Emotional feedback on timeless wavelengths,
Bearing a gift beyond price, almost free.

—*The Spirit of Radio,* Rush

They call our hobby Amateur Radio. The word "amateur" comes from the Latin "amator"—*lover*. We're lovers of radio. We're enamored of the notion of communicating over great distances without any physical connection.

Of course, the heathen nonbelievers don't understand our obsession.

"Couldn't you just pick up the telephone and *call* the people you're talking to?" Blasphemy!

They don't "get it" and probably never will. They can't appreciate the magic of sending and receiving signals with your bare hands, so to speak. The heathens only know communication as it exists through the graces of multibillion dollar corporate and government networks. They grab a telephone, punch in a number and talk. How their voices get from here to there is of little consequence to them. To us, it's *everything*!

But why do they call us "hams"? Is it because we're weird and obnoxious? Actually, the history behind the word "ham" is a little fuzzy.

"Ham" was not a complimentary term when it was first applied to our hobby. According to G. M. Dodge's *The Telegraph Instructor*, published well before the advent of radio, a "ham" was a "bad operator." As landline telegraphers became wireless telegraphers, they carried this terminology with them.

In the early days of radio, most stations communicated by generating a spark of electricity between two electrodes. *Spark-gap* transmissions were notoriously "dirty" and oc-

Gordon Fuller, WB6OVH, set up an outdoor satellite station during Field Day (an annual operating event that tests your ability to communicate during emergencies) and spoke to hams throughout the Western Hemisphere. But why go to so much trouble? Gordon could have used a telephone instead. The answer is simple: *the joy of radio*—talking to your fellow humans with a station you've assembled yourself!

The Nerd Factor

Since the earliest days of Amateur Radio, we've been tagged with an annoying stereotype: *nerds*. What's a nerd? Putting it simply, a nerd is someone of high intelligence, poor social skills and even poorer personal hygiene. The modern nerd wears clothes that are roughly 20 years behind current fashion. His hair is matted and rarely washed. Caucasian nerds tend to be quite pale in color, due to high amounts of time spent indoors. Nerds of all races tend to speak in short, clipped sentences. Their speech is occasionally accented by various facial twitches; a nerd rarely looks you in the eye when speaking. If the conversation departs from ham radio, Star Trek, the Hobbit, Dr Who, or Monty Python, the nerd usually falls silent.

Does this sound like anyone you know?

Probably not. That's the nature of stereotypes. They paint a picture with a brush that's a mile wide. I confess that I've met some hams who fit the previous description perfectly, but they are the exceptions, not the rule. It's true that Amateur Radio attracts people who are interested in science and technology. Some of these individuals are indeed "nerdish," but it's usually a byproduct of extremely high intelligence. The nerds I've met are very focused, intense people. They are completely immersed in technology, and Amateur Radio gives them an avenue to experiment and stretch their horizons.

But the majority of amateurs are as normal and well adjusted as anyone else. For most of us, Amateur Radio is not an all-consuming passion. No matter how much we love the magic of radio, our academic studies, our jobs, our friends and our families come *first*.

Ninety-nine percent of all hams defy the "nerd" stereotype. Take Dave Patterson, WB8ISZ, as an example. He is a well-rounded amateur who enjoys the hobby, but makes home and family his top priorities.

cupied large chunks of the radio spectrum. In the unregulated environment that prevailed, interference was a huge problem. Government stations, ships, coastal stations and Amateur Radio operators all slugged it out for signal supremacy in each other's receivers. Amateur stations were often very powerful, and two hams working each other across town could effectively jam all the other operations in the area. This caused the commercial guys to complain bitterly on the air about, "...THOSE #$!@ HAMS JAMMING EVERYTHING."

Amateurs, possibly unfamiliar with the real meaning of the term, picked it up and applied it to themselves. As the years passed, the original derogatory meaning disappeared. To this day, the nonbelievers may not know us as Amateur Radio operators, but say "hams" and you'll probably see a glimmer of understanding in their eyes.

"Oh, yes. It's that radio *hobby*."

No, Amateur Radio is much more. It's also the spirit of service to the community. When a tornado rips through a town and destroys all conventional means of communications, you don't hear the authorities say, "Quick! Call out the model train collectors!" In times of crisis, hams are among the first to jump into action. Not only do we have the radios, we know how to use them. We're trained communicators.

If you ever visit ARRL Headquarters in Newington, Connecticut, look for a small, granite monument that rests about 100 feet to the left of the main building entrance. On that stone you'll find the names of hams who have given their lives while using their *hobby* to save others. Do you think a similar memorial exists for bird watchers?

What an impressive station! This ham must be a millionaire, right? Wrong. John Sanford, WB8SVN, collected this gear over a period of more than 10 years. As you can see, it's easy to be fooled by awesome photographs.

IS HAM RADIO EXPENSIVE?

When you're taking your first steps into a new venture, it's wise to keep your eyes wide open. Unfortunately, many Amateur Radio books and articles try to minimize the issue of cost, or avoid it altogether. The authors have the best intentions. They're worried about scaring you away from the hobby.

I'd like to take a different approach with this book. I believe that you're mature enough to understand that there are some substantial costs involved. On the other hand, Amateur Radio is not the most expensive hobby in the world. If you want to see how bad it can *really* get, talk to a private pilot who owns his or her own airplane. General aviation makes hamming look like rock collecting by comparison.

There is a decent chance that you'll fork over at least $200 to get started in Amateur Radio, probably more. Two hundred dollars is a lot of money for many people. (It annoys me how some authors attempt to trivialize the impact of "a few hundred dollars.")

You'll probably hear that you can save huge amounts of money by building your own gear from kits. This is true to a degree. Most transceiver kits are low-power HF rigs, although you'll find the occasional VHF or UHF kit. If you have sufficient technical experience to build a kit and fix it if something goes wrong, a kit is a viable cash-saving option.

But if you want a transceiver with sophisticated features and higher power output, and if your engineering skills are lacking, you're looking at new or used equipment. Now we're talking about potentially large price tags.

The good news is that you won't be shelling out the big bucks on a frequent basis. A new dual-band FM mobile transceiver can set you back about $700, but you probably won't buy another one for a *long* time. If you take good care of a transceiver, it can theoretically last a lifetime. There are hams today who are using transceivers manufactured in the 1950s. Those radios work just as well today as they did when they were new.

Set reasonable goals for yourself and you can save even more money. For example, forget about tall towers and huge antennas for now. They're *very* expensive. Instead, fix your sights on less ambitious antennas that you can install on your roof, patio or wherever. If your passion is to operate on the HF bands, consider wire antennas (see Chapter 7). You'll have a blast with a wire *dipole*, at a cost of about $25 or less.

When you see those photographs of elaborate, wall-to-wall ham stations and forests of steel towers, consider two things:

(1) **The ham in question might be rich.** There's nothing wrong with wealth, but most of us don't have that advantage at the moment.

(2) **The gear may have been accumulated over a long period of time**. This is what usually happens. You buy a radio. A couple of months or years later, you buy another. Perhaps you receive a couple of accessories as gifts. After a decade of purchasing and gift giving, you wind up with a station that would make the Voice of America jealous!

THE SOCIAL SIDE OF HAM RADIO

Hamming is much more than hardware. In the days before political correctness expunged sexist language from our vocabularies, Amateur Radio was called a "fraternity." Although the word has fallen out of fashion, the idea is still valid. There is a

Should I Join a Club?

Some clubs are extremely active. Their members get together for contests, transmitter hunting, Field Day, and various public-service events. Many of these clubs maintain stations that you can operate.

Unless you're in desperate need of mind-numbing gab sessions, you should obviously seek out an active group. By joining a club and becoming involved in the club's projects, you'll make new friends very quickly. You'll also have a pool of veteran hams that you can tap for answers to tough questions. Many of them may even be willing to come to your home to diagnose a problem, or help you erect a new antenna.

To locate the ARRL-affiliated club nearest you, call 1-800-32-NEW-HAM. Ask for a list of clubs in your area. Not only is the list free, so is the call.

bond between hams that defies an easy description. Taking my best shot, I'd say the bond is similar to what you'd find among members of organizations such as the Freemasons, Knights of Columbus, Odd Fellows and so on. It's the idea that we hams share "secret" knowledge that is beyond the understanding of the general public. And, to a great extent, that's true! When you consider the fact that most people can't set the digital clocks on their VCRs, they must think we're magicians!

That fraternal bond manifests itself in many ways. You might be talking to someone at a party and casually mention that you're a ham. "Really? So am I! I'm KD7XZY!" You give him your call sign and it's as though you've just exchanged a secret handshake.

Perhaps you're on vacation and you've just accessed a repeater in a strange town. You ask for directions to a restaurant. Someone answers your call, gives you the directions . . . then asks if you'd like to meet for breakfast the following morning!

Like many hams, I used to have my call sign on the license plates of my car. Once I was driving through eastern Montana. There was nothing to see but rolling fields of wheat. I seemed to be the only traveler on the road. Soon another car appeared behind me. As it drew closer, the driver began to honk. I had no idea what was going on until, to my astonishment, I realized that he was honking in Morse code, sending my call sign over and over!

For the next 20 minutes we cruised down the interstate, honking to each other like lunatics. He asked if I had a 2-meter FM radio. No, I replied. So he sent his call sign, told me his name was Ralph, and said that he was on his way to Miles City, Montana. I pounded the horn switch, sending my particulars in sloppy CW. After a while he said that he had to exit for gas and ended with a snappy "73" (best wishes). That was, without a doubt, the most unusual CW conversation I've ever enjoyed! Can you think of any other hobby that would inspire a scene like that?

Hamfests

When hams feel the need to meet each other face-to-face, they often do it at gatherings known as *hamfests*. Even the smallest hamfests usually include a flea market where hams sell and trade used equipment. If you're in the mood to spend money on

Even a small hamfest, like this one in Quincy, Illinois, usually has a flea market where you can find excellent used-equipment bargains.

The dickering is fast and furious at the Dayton *HamVention* flea market, the largest of its kind in the world!

preowned gear, a hamfest is the place to do it. The larger hamfests will also attract new-equipment dealers, and the *very* largest will have exhibits from the manufacturers themselves.

Medium to large-size hamfests will often offer forums where you can take part in discussions on various topics. Some forum topics are technical while others are political. Many hamfests also have food available on the premises, and even entertainment.

No matter where you live, there is probably at least one hamfest each year within 100 miles of your location. If you live in the more populated regions of our country, you can count on having a dozen hamfests or more within a comfortable driving distance. Check the Hamfest Calendar in each issue of *QST* magazine to see what's coming up in your area. If you operate packet radio, you'll also find hamfest announcements on your local BBS. Your local club may sponsor a hamfest. If not, it's a safe bet that someone in the organization knows when the nearest hamfest takes place.

But just as Moslems are urged to make at least one pilgrimage to the holy city of Mecca, you must make at least one journey to the largest hamfest on the face of the planet: the Dayton *HamVention*. Each year in late May, 40,000 Amateur Radio operators swarm into Dayton, Ohio, and converge on the Hara Arena convention center. The result is a gigantic ham radio flea market, along with the most extensive collection of Amateur Radio dealers and manufacturers in the world. If you can't find it at the Dayton HamVention, it may not exist!

The three-day event draws a vast cross section of the human species. Some make the pilgrimage to sell, others to buy. They march up and down the endless aisles of the outdoor flea market and cram themselves into the arena complex. The air is so full of radio signals (mostly from FM hand-held transceivers), interference is horrendous. Cars honk unceasingly. Television news helicopters hover overhead. The odor of hot dogs, beer, cigarettes and soft drinks wafts through the hallways. If you're lucky, it won't rain . . . or at least the tow trucks will be able to rescue your car from the mud if it does.

Why would anyone subject themselves to such a spectacle? It's that darn fraternal notion again. As you shuffle wearily through the crowds you can't help catching the glances of equally weary souls. That millisecond meeting of the eyes says it all: "Greetings, fellow magician. We're suffering, but we're here. Isn't it grand?"

Are Codeless Technicians "Real" Hams?

When the FCC created the codeless Technician license in 1991, it caused an explosion of controversy that reverberates to the present day. Morse code is tightly interwoven with the traditions of Amateur Radio. For some hams, the idea of earning an Amateur Radio license without passing a code test is highly offensive. They feel that you aren't a true ham until you can demonstrate mastery of Morse code.

Fortunately, these cantankerous hams are a dwindling minority. The codeless Technicians have more than proven their value to Amateur Radio. The license has also become the most popular ticket into our hobby. If you're a new ham, the odds are you're a codeless Technician!

As far as the American Radio Relay League is concerned, codeless Technicians are first-class hams—even if they never pick up a code key as long as they live. Codeless Technicians are particularly important for the trailblazing work many of them are doing on the UHF and microwave bands. We'll lose these bands soon if we don't put them to good use. Thanks to codeless Technicians, we stand a chance of preserving our high-end spectrum!

If you hear someone declare that a codeless Technician isn't a real ham, ignore the insult. (What is a "real" ham anyway?) Their opinion is grossly out of date and not likely to change. Arguing with them is like attempting to teach a pig to sing—it wastes your time and annoys the pig.

DON'T SIT STILL

Books are great things, but don't keep your nose buried in one too long! You won't savor the spirit of radio by flipping pages.

This book is *not* the be-all end-all reference for Amateur Radio. You won't, for example, find construction projects in this book. Pick up a copy of *The ARRL Handbook* or *The ARRL Antenna Book* and you'll find a lifetime's worth of projects. And if you need more detail on every Amateur Radio communication mode and how to use it, buy *The ARRL Operating Manual*. These books are available from your favorite dealer or directly from ARRL Headquarters.

The purpose of this tome is to get you on the air as quickly and painlessly as possible. You'll learn which hardware combinations work best, and how to put them together to create a station for a particular mode. You'll also receive valuable operating information to help you avoid the "newbie" label!

Browse the chapters and decide which activities interest you the most. (The chapters are organized according to the most popular Amateur Radio pastimes.) Then, read your chosen chapters carefully. You'll find that the reading is easy and even fun.

After that, *put the book down*, get on the air and start enjoying Amateur Radio! It's okay to reread the book to refresh your memory, or when it's time to explore a new facet of our hobby, but there's no substitute for the total joy of communicating without wires. The sooner you start burning the air with electromagnetic waves of your own creation, the sooner you'll appreciate the magic!

FM—No Static At All

Nothing but blues and Elvis,
And somebody else's favorite song . . .

—from "FM" by Steely Dan

You won't hear blues or Elvis on amateur FM—unless the late King of Rock 'n Roll decides to get his ham license in the afterlife. No, FM as we know it in Amateur Radio is another animal entirely. (Many hams will tell you that there *are* some similarities!)

When you hear the word "FM" you probably think of a station that blares the hits somewhere between 88 and 108 on your radio dial. The music is crystal clear and there are few, if any, annoying buzzes or other noises. As the song says, "No static at all."

But FM actually stands for *frequency modulation*. It's a method of transmitting information that involves shifting the frequency of a radio signal back and forth in sync with voices, music or whatever. (The amount of frequency shift is known as *deviation*.)

AM means *amplitude modulation*. It's almost the reverse of FM. An AM transmission is comprised of a *carrier* signal and two *sidebands*. You send information on AM by shifting the *strength* (amplitude) of the sidebands. Unlike FM, however, the frequency of an AM signal never changes.

If you built a radio that listened *only* for signals that shifted their frequencies, you wouldn't hear AM signals at all, would you? And since most of the static in the world is amplitude modulated (Mother Nature must be fond of AM), your clever radio would automatically reject noise! Congratulations. You've just designed an FM receiver. (You're about 75 years too late, but who's counting?)

As you've probably guessed by now, FM transmissions can take place on *any* frequency, not just those between 88 and 108 MHz. It's federal law, courtesy of our friends at the Federal Communications Commission (FCC), that dictates where FM signals can appear. Hams are allowed to transmit FM mainly on our VHF and UHF frequencies (those above 50 MHz). We also have a small FM segment between 29.5 and 30 MHz.

Why FM?

When it comes to clear two-way communication, FM is way ahead of AM. It's a pleasure to cruise the highways and chat with your buddies without noise interference. You can be driving through the granddaddy of all thunderstorms and hardly hear a peep of lightning static in your radio. Try the same thing with an AM transceiver and you'd end up deaf, insane or both.

Assuming that your radio is connected to a decent speaker, FM audio has wonderful fidelity. If your friend is speaking clearly (as opposed to screaming and cursing), you'll hear every word. During some conversations it sounds as though the person is right there in the car with you—whether you like it or not.

The only skunk at the party is *range*. Unlike shortwave signals, VHF and UHF transmissions rarely go bouncing off the upper layers of the atmosphere to land on other continents. Instead, they take the straight-line express route to outer space. This means that your communication range is limited to a short distance beyond your local horizon *at best*. (Yes, there are exceptions. We'll talk about those in later chapters.) So, to cover a worthwhile amount of territory on VHF and UHF, you need a little help.

ENTER THE FM REPEATER

In simplest terms, a repeater *repeats*. It's an electronic parrot. A repeater listens to a signal and repeats what it hears. Unlike a parrot, however, a repeater repeats the signal *while it's listening to it*.

From a technical point of view, a repeater is a station that operates automatically. There is an antenna, a transmitter and a receiver, but no human operator pressing buttons and flipping switches. A microprocessor-based device known as a *controller* is the brains of the outfit. The controller makes sure everything operates properly, or takes the repeater off the air if it doesn't.

The repeater's receiver is exquisitely sensitive, and its transmitter usually operates at high power, typically 100 W or more. The antenna is an *omnidirectional* type. (In other words, it receives and transmits in all directions.) If you place a repeater system on a mountaintop, hill, skyscraper or radio tower, it can receive and transmit over an enormous distance—even on VHF or UHF.

Take a look at Figure 2-1. Let's say your puny mobile radio can only cover a few miles on the ground. You don't have a snowball's chance of contacting someone, say, 20 miles away. But the repeater, from its lofty perch, hears both you *and* the other fellow.

The instant you begin speaking, the repeater receives your signal and blasts it out over hundreds or thousands of square miles. You keep talking and it keeps repeating. Thanks to the repeater, your distant buddy hears your melodious voice booming in his radio. Of course, he isn't picking up your signal directly. He's listening to the transmissions of the repeater.

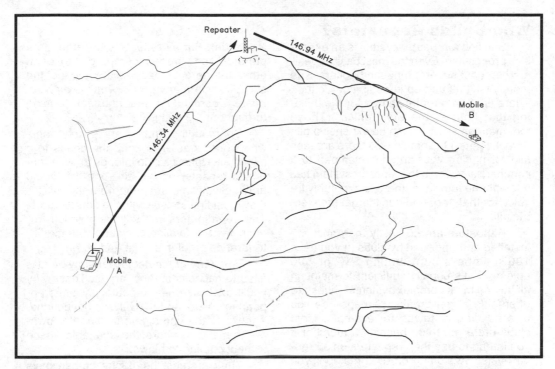

Figure 2-1—The mountain blocks direct communication between mobile stations A and B. When mobile A transmits, however, the repeater hears its signals and relays it to mobile B.

Only one person can use a repeater at a time. When you release your microphone button, it's your friend's turn to talk. The repeater picks up his signal and relays it back to you. Once again, you're not hearing him directly. You're listening to the repeater acting as the middleman in your conversation.

One of the keys to this technological miracle is the repeater's ability to receive and transmit simultaneously. This may seem like a simple trick, but it isn't.

Separate Frequencies, Separate Energies

Have you ever watched someone use a public address system that's horribly out of adjustment? Some poor soul strolls up to the microphone, clears his throat and whispers, "Is this thing on?"

The amps are usually turned up so high that the sound of his amplified voice is picked up by the microphone. The PA system quickly amplifies the voice again, and again, and again until a banshee howl fills the auditorium and sends everyone diving under their seats. This painful phenomenon is known as *feedback*.

Feedback can happen in radio devices just as easily, and the consequences can be just as awful. If a repeater transmitted and received simultaneously on the same frequency, you'd have a horrendous feedback loop. Yes, the repeater would howl, too!

Who Builds Repeaters?

Building a repeater system is an expensive proposition. Even the most basic repeaters can cost several thousand dollars. The money isn't all tied up in equipment, either. There is often a rental charge for the building that houses the electronics. (Those mountaintop shelters can be expensive bits of real estate!) In some cases there are fees just for having your antenna clamped to a commercial tower. If the repeater system has a telephone interface, there is a monthly invoice for that, too. And don't forget the electric bill!

Repeaters are so costly that most are installed and operated by clubs. If you have 100 members contributing $10 or $20 apiece, that's usually sufficient to maintain all but the most complex systems. Clubs are often able to get breaks; perhaps free use of a building or tower, for example. Most clubs prefer that you become a member if you intend to use their repeaters on a regular basis. By signing on as a member you may receive "secret" codes that allow you to operate some of the repeater's special functions—such as the autopatch.

Can Anyone Put a Repeater on the Air?

Back in the gold rush days of amateur FM, repeaters popped on the air like mushrooms after a summer rain. The atmosphere was totally Dodge City—little law and even less order. But as repeaters became more popular, they began bumping into each other. Interference became the rule, not the exception, in many areas. That's when *repeater coordinating groups* began forming throughout the country.

Coordinating groups are simply hams who work together to keep the chaos to a minimum. Unless a club intends to put a renegade repeater on the air—one that could draw unwelcome attention from the FCC— they must first consult with the coordinators. The coordinators will tell them which frequencies are available, where potential interference problems exist and so on.

The 2-meter amateur frequency band is filled to capacity in most areas. That's why you'll often see new repeater systems appearing on 222 and 440 MHz, if not beyond. Thanks to the coordinators, repeater owners manage to share the available space without coming to blows.

Their greatest headaches these days are caused by dual-band VHF/UHF radios that have the ability to function as crossband repeaters. They can transmit a 440-MHz signal on 2 meters, or vice versa. Some hams are throwing them on the air without considering the possible conflicts. If you own one of these electronic wonders, *think* before using the crossband feature. The best advice is to use it sparingly, if at all.

So what's a poor repeater to do? The answer is to transmit and receive on *separate* frequencies. Let's say we have a repeater that listens for signals on 147.96 MHz and repeats whatever it hears on 147.36 MHz (see Figure 2-2). Notice that the input and output frequencies are separated by 600 kHz. This is the standard for 2-meter repeaters, although you'll occasionally find repeaters using different *splits* or *offsets*.

On 50 MHz the standard offset is 500 kHz. One MHz is the common split on 222 MHz and a 1.6-MHz separation is the plan at 440 MHz. There are exceptions to this rule, as experience will teach you.

But just when you thought all these heady concepts were falling into place, here's

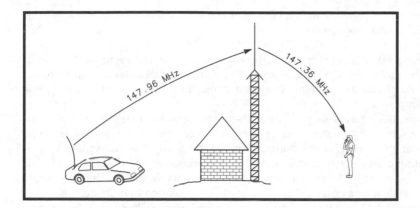

Figure 2-2—Repeaters use a split-frequency scheme to avoid nasty feedback problems. In this case, the repeater receives the mobile operator on 147.96 MHz and transmits what he is saying on 147.36 MHz.

An inside view of the W1KKF repeater in Wallingford, Connecticut. The controller takes center stage. The duplexers are barely visible at the bottom of the photograph.

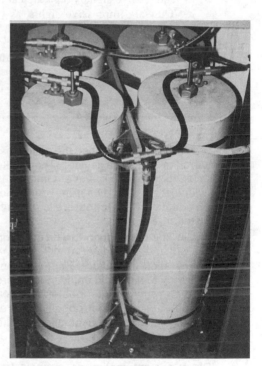

Figure 2-3—These strange-looking cylinders are duplexers. They keep the received signals and transmitted energy from mixing in destructive ways!

a new curve ball for you. Not only do repeaters transmit and receive at the same time, most do it *on the same antenna!*

Say *what*?

You heard right. Crazy as it seems, there's no law against receiving and transmitting simultaneously with the same antenna. The incoming and outgoing waves of energy pass like ghosts in the night. They don't smash into each other like a cosmic freeway pileup.

The trick, however, is keeping the high-power transmit energy from finding its way into the sensitive repeater receiver. A repeater receiver is designed to deal with signals carrying the energies of a butterfly sneeze. If the repeater transmitter was allowed to dump even a fraction of its power into the receiver, the result would be the equivalent of dropping a 50-pound cinder block on a Sony Walkman. Not only would the receiver be deaf as a post, its life span would be measured in milliseconds.

The magical device that keeps the receiver from being cooked to a golden brown is known as a *duplexer*. Duplexers are extremely sharp filters that keep the transmit and receive signals separated at the point where they enter or exit the antenna system. They are impressive sights, as you can see in Figure 2-3. They look like big metal cylinders. Duplexers for 2-meter repeaters can be more than three feet long. Six-meter repeater duplexers are true monsters with some as tall as six feet or more.

Be careful not to confuse a duplexer with a *diplexer*. Diplexers (and *triplexers*) are

FM Without Repeaters

We've clearly established the notion that repeaters are wondrous devices. They make it possible for you to talk over a humongous area with just a watt or two from your radio. But the equation **FM=Repeaters** is *not* written in stone!

If your signal will span the distance on its own, why use a repeater at all? You can talk from station to station—from your antenna directly to his—and enjoy much more privacy than you do on a repeater. And you don't need to worry about hogging the repeater with your endless observations on the future of mankind, the hottest gossip, or whatever.

Direct communication on a single frequency is known as *simplex*. It simply means that you're communicating in one direction—and one direction *only*—at any given time. *Duplex*, by comparison, means that you're communicating in both directions simultaneously. Telephones are duplex devices because you can speak and listen at the same time.

There are FM simplex frequencies set aside on every amateur VHF and UHF band. If you're looking for random contacts, however, keep your ears tuned to the national FM simplex calling frequencies. They're shown in the sidebar, "Where's The Action?" Of all the national calling frequencies, the most popular is 146.52 MHz. If you hear someone say, "Let's go simplex," or "Let's go to 5-2," they usually mean 146.52 MHz.

If you live in or near a large urban area, don't use the national simplex calling frequencies for long conversations. There's a decent chance that you'll run into interference from other stations.

The rule of thumb is to use simplex whenever possible. Keep the repeater free for hams who really need it. If you can talk to the other station without the repeater's help, why not let someone else use it?

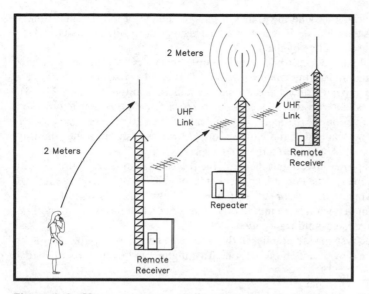

Figure 2-4—Many repeater systems use more than one receiver to provide reliable coverage over a wide area.

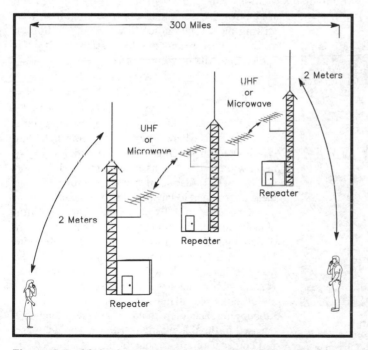

Figure 2-5—Linked repeater systems allow FM-active hams to cover huge areas—sometimes several states at once—with nothing more than an H-T.

used to separate transmit and receive signals when they've at vastly different frequencies (usually different bands altogether). They're compact devices, intended for use with dual- and tri-band VHF/UHF transceivers. They allow you to operate these rigs on multiple bands with a single antenna.

When it comes to separating signals that are much closer in frequency, only a duplexer will do. Unfortunately, you'll often see diplexers referred to as duplexers in advertisements and catalogs. The difference is easy to spot, though. True duplexers cost $1000 or more, and you can't fit them under your dashboard!

Let's Get Complicated

So far we've been talking about simple repeater systems; one transmitter, one receiver and one antenna. Repeaters can become much more complicated, however.

Figure 2-4 illustrates a receiver system that has one transmitter, but many receivers. This is a great way to fill the "holes" in a repeater's coverage. Each receiver sends whatever it hears back to the main site via a UHF or microwave link. At the site there is a device that analyzes all the incoming signals and selects the strongest one to pass to the repeater transmitter.

You can usually tell when you're listening to a repeater system that's using remote receivers. If someone is carrying on a conversation while driving around town, they'll pass in and out of the coverage area of various receivers. As they do, you'll hear their signal become noisy, then switch to crystal clarity. (Cell phone users will recognize this!)

Repeater complexity doesn't end with multiple receivers. A repeater can have multiple receivers and transmitters *on other bands*. This allows *crossband linking*. For example, a 2-meter repeater may have a receiver and transmitter on the 222-MHz band so that Novices can join the 2-meter operators. The same repeater may have a transmitter and receiver on the 10-meter FM subband. When 10 meters opens up, everyone on the system—including the Novices on 222 MHz—can jump into the fray. (Before you ask, yes, this is all perfectly legal.)

The ultimate repeater coverage scheme is the *linked* system (see Figure 2-5). In this type of system, many repeaters are connected through UHF or microwave links to create a huge network that can spread over hundreds or thousands of miles. On a linked system, a ham in one city can talk to other hams in distant cities—with nothing more powerful than a hand-held transceiver.

Linked repeater systems are popular in the Midwest and Far West. The Evergreen Intertie, for example, covers much of Oregon, Washington, and even parts of British Columbia.

YOUR OWN FM RADIO

In the amateur FM universe a "station" can be a tiny hand-held transceiver you wear on your belt. It can also be an electronic assemblage that occupies half your writing desk. The choice is yours, depending on your cash flow and the temperament of your spouse, parents or whomever.

FM operating on the road doesn't necessarily involve a car! Russell Dwarshuis, KB8U, enjoys his H-T from a recumbent bicycle near Mt Ashland, California.

H-Ts

The hand-held transceiver, or *H-T*, is the trademark of the FM operator. If you go to a hamfest you'll see them everywhere. They dangle from belts like six-shooters in an Old West movie, although I've never witnessed two hams engaging in an H-T shoot-out. Many H-Ts also sport speaker/microphones and these are clipped to lapels and shirt pockets.

Of course, an H-T wouldn't be a true radio without the ubiquitous *rubber duck* antenna. If you're old enough to remember the CB craze of the mid-'70s, you no doubt recall C. W. McCall's languid drawl, "This here's the Rubber Duck, come on." Well . . . this rubber duck has nothing to do with that god-awful song! (Not all music memories are good!) The rubber reference refers to the fact that the antenna is encased in flexible plastic or rubber. The duck part just comes naturally to some folks.

H-Ts are wonders of miniaturization. The manufacturers pack everything but the proverbial kitchen sink into these radios. Your typical H-T puts out about 2 W of RF power, enough to pack a solid signal into a sensitive repeater. Some will produce even more power with optional battery packs. *DTMF* (*Dual-Tone Multi Frequency*) telephone-style keypads are standard equipment, as are bright liquid-crystal displays (LCDs). Most H-Ts also offer extended receive capability (listen to more than just hams!), paging (bug your buddies anytime, anywhere), scanning (never miss a single signal) and much more.

The downside of all these goodies is that modern H-Ts are notoriously complicated. Learning to simply program a repeater channel could require a substantial chunk of time. If you want to delve into the netherworld of paging or other arcane magic, set aside a millennium or two. (Just kidding. A decade ought to do the trick.)

H-Ts are available in single-band models, dual banders and—for the ultimate in mind-bending complexity—*tri*banders. Obviously, the more bands the radio offers, the deeper you'll need to dig into your checking account. If most of the activity in your area is on the 2-meter band, save your cash and buy a single-band radio. If you live in a densely populated region, you're more likely to have large numbers of repeaters on more than one band. In this situation a dual-band H-T makes sense. Even a tribander might be worth the investment in some areas. Before you

A virtual cornucopia of dual-band H-Ts! These hand-held transceivers include every feature imaginable . . . and then some!

reach for your wallet, check *The ARRL Repeater Directory* and determine what bands are most active where you live. Don't buy more radio than you need!

So, should you run out and purchase your own H-T? Here are a few guidelines . . .

Buy an H-T if:
* You plan to operate primarily in your car, on foot, or on a bicycle. An H-T's portability is terrific for hams on the go.

- You have at least one powerful repeater in your area. With the low output power of an H-T, you need a repeater that can relay your signals reliably.
- You plan on doing a fair amount of public-service work. H-Ts are almost mandatory for public-service activities these days. When you're providing communication for a foot race, for example, you don't want to be tied down to your car.

Base/Mobile Transceivers

Like H-Ts, mobile transceivers have benefited from the revolution in miniaturization. The typical FM mobile radio of the '90s is probably smaller than the AM/FM broadcast radio in your dashboard. Even so, most pack a big RF punch with output power ranging from 30 to 50 W. That's more than enough "oomph" to hit all but the most distant repeaters.

Mobile rigs are also packed with every feature imaginable. A mobile radio can usually do anything an H-T can, and more! Despite their small sizes, mobile rigs are easier to use than H-Ts. The buttons and other controls are farther apart, making it possible for even the chubbiest fingers to use them. The LCD displays are also larger. A larger display is much easier to read.

Generally speaking, the receive audio quality of a mobile radio is superior to that of an H-T. Mobile speakers are larger and the case offers at least some acoustic baffling. Audio clarity and power are very important for mobile operating. When the wind is whistling through the open window and your favorite song is blaring at the threshold of pain, you still want to be able to hear your radio, don't you?

This Yaesu FT-5100 is typical of dual-band mobile transceivers. It offers 50-W output on 2 meters and 35 W on 440 MHz.

A mobile radio installation is clean and convenient. (Assuming you don't go after your dashboard with a chain saw and a belt sander.) You can install the radio in the dash, or under it. The antenna cable connects to the back panel, out of sight. The same is true of the dc power cable. Many mobile transceivers are designed to mount on rails that allow you to slide the rig out and take it with you when you leave the car. A few can even be installed in the trunk with just a tiny control panel and microphone connector in your dashboard. (Imagine the consternation of the thief who steals that control panel. "Gee, this dang radio don't work! I wonder why?")

Mobile rigs can also do double duty as *base* radios in homes, apartments, offices or wherever. When a mobile transceiver is in your car, it's drawing power from the automotive battery. In your home, you'll need to provide the same dc power. You do this through a device known as a *power supply*. A power supply takes the 120 V ac from the wall outlet and converts it to 12 V dc for the radio.

There's No Such Thing as Too Much Current

The key to buying a power supply is getting the right *current capacity*. Your mobile transceiver draws the highest amount of current when you're transmitting. So, the power supply must be capable of providing *at least* that much current on a *continuous basis*. Beware of ads for high-current power supplies that seem incredibly inexpensive. Read the specifications. If the power supply is rated at 20 A, is that continuous or intermittent (often called "ICS" or "surge")? The intermittent current rating will always be much higher than the continuous rating. If your radio needs 20 A of current when you're transmitting, your power supply must be able to supply 20 A *continuously*. Don't worry about buying a power supply that offers more current than you need. Your radio will draw only as much as it requires. If you can find a good deal on a power supply that offers two or three times as much juice as you need, go for it! You can always use the extra current capacity to power other goodies in the future.

Used or New?

Unless you're a gambler, avoid *large* investments in used equipment—radios or otherwise. The exception is when you can buy your gear from a reputable dealer who'll stand behind it for at least 30 days. Sometimes it takes that long for problems to show up.

If you're a whiz at fixing electronic circuitry, or know a good friend who is, you're in a different category. There are plenty of used-equipment bargains out there. Check out your nearest hamfest flea market, or the "Ham Ads" classified section of *QST* magazine, and I bet you'll find plenty of used FM radios for very reasonable prices. Some of these radios are great. You fire them up and they perform like new rigs. Others are . . . well . . . disappointing.

Hams have excellent reputations as honest peddlers of second-hand gear. Even so, there are exceptions to the rule. I've been told that something was working "just fine," only to discover that the seller was exaggerating quite a bit! Once I had to throw away a large item that I'd purchased at a hamfest. After trying to make it work, I discovered that it was damaged beyond the cost of fixing. I've also put in some very long hours repairing other "bargains."

If you have any doubt about your technical abilities, stick with new radios. You'll enjoy the advantage of the manufacturer's warranty. If the rig gets sick during the warranty period, it's not your problem.

ANTENNAS

The best radio in the world isn't worth a stale donut without a good antenna. Imagine a stereo system with a $3000 amplifier and $20 speakers. How do you think it will sound? Probably like great audio being mangled by cheap speakers!

Antennas don't affect the way you sound, except in the sense that your signal might be pretty noisy on the receiving end if your antenna is poor. The type of antenna you choose determines how far you can talk. In other words, the better your antenna, the greater your *range*.

H-T Antennas

Thanks to repeaters, you can get away with using some truly awful antennas and still cover a decent amount of turf. The sensitivity of the repeater makes up for your lousy signal. That's why the rubber duck antennas supplied with every H-T appear to give adequate performance. In truth, they're not very good antennas, but they have the distinction of being flexible and virtually unbreakable.

Rubber ducks are fine if you expect to always be within range of a powerful repeater system. If you think you'll find yourself on the fringes, however, consider

Home Antenna Installation Tips

- Solder your coax connectors *carefully*. Too little heat will result in a poor connection, but too much heat can damage the cable. See Figures 2-12, 2-13 and 2-14 for step-by-step instructions on soldering three popular connectors. Get a veteran ham to help if you're in doubt.

- Buy an accurate VHF/UHF SWR meter and learn to use it!

- Weatherproof the coax connector after you attach it to the antenna. Cover it with a commercially available putty compound such as *Coax Seal*, or coat the connector with silicon grease and wrap the entire assembly in several tight layers of electrical tape

- If you intend to use a chimney as an antenna support, inspect the chimney first. Make sure the bricks are still firmly in place. Even small antennas act like sails in a strong breeze. They'll add structural stress to your chimney.

- Be careful when creating a hole in your roof, siding or wherever for your coax. Make sure you're not about to drill into electrical lines or plumbing. Use generous amounts of silicon caulk to seal the hole once the cable is in place. Don't forget to add a *drip loop* at the point where the cable enters the house. A small U-shaped loop in the cable will keep rain from seeping into your home.

- When installing the coax, don't attempt sharp bends that could crimp the cable. This will change the impedance of the cable at the crimp and cause a mismatch. By the same token, take care not to crush the coax with fasteners, clamps and so on.

- If you're putting up a beam antenna, be sure to calibrate the direction the antenna is pointing with the indication on the controller in your home. (Having a friend available during this step is a big help.) When the controller says that your antenna is pointing west, for example, you want to know you can trust it!

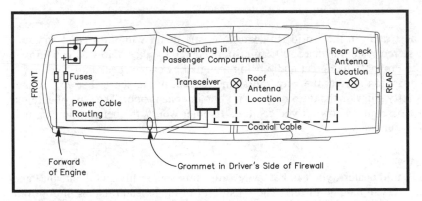

Figure 2-6—When installing a mobile antenna, route the coax to your radio by slipping it behind seats and under carpets. At the same time, keep the length as short as possible. If you're using a mobile transceiver (as opposed to a hand-held), attach the power cable directly to the car battery and put fuses in both leads.

investing in one of the many telescoping $\frac{1}{2}$- or $\frac{5}{8}$-wavelength models you'll see advertised in *QST* magazine. They'll give you noticeably better coverage.

Mobile Antennas

An external antenna is the key to successful mobile operating. Yes, you can use an H-T inside a car with its rubber-duck antenna, but the results are often poor. Put a good antenna on your car and you'll likely *double* the range of your H-T! Even 50-W radios need good mobile antennas to get the best performance possible.

The antenna connects to your radio through a length of *coaxial cable* (see Figure 2-6). The type of coax you use for mobile applications isn't very critical because the length is so short. It *does* make a difference in home antennas, however, as you'll see later.

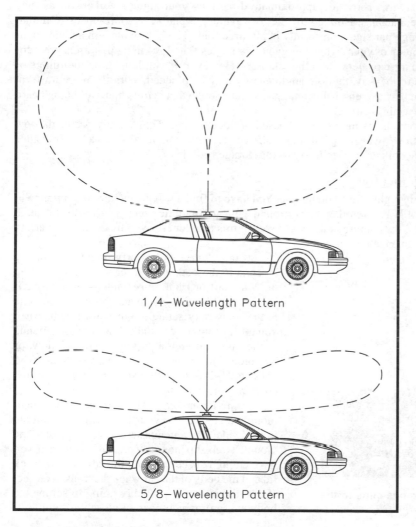

1/4—Wavelength Pattern

5/8—Wavelength Pattern

Figure 2-7—The real difference between a $\frac{1}{4}$- and $\frac{5}{8}$-wavelength mobile antenna involves the radiation patterns. A $\frac{1}{4}$-wavelength antenna has a more-or-less uniform pattern. A $\frac{5}{8}$-wavelength antenna radiates your signal in a pattern that sends the energy toward the horizon.

If you're worried about the aesthetics of your automobile, you can relax a bit. When it comes to VHF and UHF mobile antennas, we're not talking about a 15-foot monstrosity strapped to your bumper. Most FM mobile antennas are fairly short, typically less than a few feet long. (Mobile antennas for 440 MHz are just *inches* long.) You'll find mobile antennas for single bands or multiple bands. The 2-meter/440-MHz antennas are especially popular among owners of dual-band radios.

The Longer the Better?

As you browse the mobile antenna ads you'll see references to $1/4$, $1/2$ and $5/8$-wavelength antennas. The greater the fraction, the longer the antenna. Most hams assume that a $5/8$-wavelength antenna would perform better than a $1/4$-wavelength antenna. This is true to a certain extent, but it depends on your operating environment.

Figure 2-7 shows the radiation patterns for typical $1/4$- and $5/8$-wavelength antennas. The patterns show the approximate directions your signals will travel as they leave these antennas. Notice how the $1/4$-wavelength pattern is more-or-less circular. It's radiating your signal uniformly in all directions. The $5/8$-wavelength antenna concentrates much of your signal energy at low angles. The low-angle pattern is great for hitting distant repeaters and other stations. However, if you have hills, buildings or mountains in the way, the low-angle energy may be blocked. With the more uniform pattern of a $1/4$-wavelength antenna, you stand a somewhat better chance of being heard when the terrain is rugged.

For some hams the choice is based on looks alone. They simply prefer the low profile of the $1/4$-wavelength design. A $1/4$-wavelength antenna also has a better chance of avoiding collisions with low parking-garage roofs!

Attaching Antennas to your Car

No, super glue won't do the job. You have to find a way to anchor your antenna in the 55+ MPH breezes that blow around your automobile when you're on the road. Fortunately, the manufacturers have solved this problem for you by creating a variety of mobile antenna mounts . . .

COMET ANTENNAS

This little 440-MHz antenna clips to the gutter just above the window.

Mag mounts—A powerful magnet holds the antenna to the body of your car. You can set it up on your roof or trunk in seconds—and remove it just as quickly. The disadvantage of mag mounts is that they may scratch your car's finish. Also, you must bring the coaxial cable from the antenna to the radio through a partially open window, or by some other means. (I've heard of hams who get away with simply closing a car door on the coax, but this isn't the best approach.)

Trunk-lip mounts—These are semipermanent mounts that attach to your trunk lid, often on the hinged side. The antenna screws or snaps onto the mount. Trunk-lip mounts work fine, as long as you have enough space between the edge of the trunk lid and the body of the car when the trunk is closed. If not, the mount is probably going to scrape the body every time you open or close the trunk.

COMET ANTENNAS

The classic magnetic mount mobile antenna. The base is nothing more than a powerful magnet that holds the antenna securely to the car. Note how the coax has to squeeze under the truck lid to enter the car.

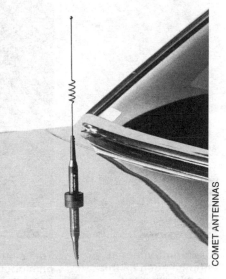

COMET ANTENNAS

If you don't mind drilling large holes in your car, this is one of the cleanest installations you can manage.

Body mounts—There's nothing complicated about these. You simply drill a large hole in the body of your precious, expensive automobile and install the antenna permanently. (I can hear some of you shuddering at the mere *thought* of doing this.) In terms of achieving a durable installation, this is clearly the best way to go. It does tend to reduce the resale value of your car, though.

On-glass mounts—These are actually complete antenna systems, not just mounts. On-glass antennas are *tre chic* among the cellular telephone set. You're simply not hip if you don't have an on-glass antenna. The ham versions look—and work—the same. The RF energy passes from one metal plate to another *through* the window glass. Although they perform reasonably well, they can be tricky devils to install. The situation is complicated further if you own a car with a window defroster that uses a thin metallic membrane in the glass. RF energy doesn't pass through this stuff very well!

Home Antennas

Do you want to radiate your home signal in all directions at once? If your answer is "yes," you want an omnidirectional antenna. These antennas are popular on VHF and UHF because they're easy to install and require little, if any, adjustment. If you install an omni antenna as high as possible, you'll enjoy plenty of contacts.

Popular omnidirectional base designs include the *groundplane*, *discone* and *J-pole*. Like mobile antennas they are available in single and multiband versions. Most VHF/UHF omni designs are relatively short, but others tower to heights of 15 to 20

An omnidirectional antenna such as this one will give you decent coverage in all directions. Note how the antenna (a *colinear ground plane*) attaches to a short mast, which is secured to the chimney with stainless-steel straps.

This multiband omnidirectional antenna is known as a *discone*.

feet. It all depends on the frequency (the lower the frequency, the longer the antenna) and the design.

There are times when you may need to focus your signal energy, particularly when you need to reach a distant station. You can't do that very well with an omni-directional antenna. Instead, you need a *beam* design. The most popular type of beam antenna for VHF and UHF is the *Yagi* (see Figure 2-8). The runner-up is the *quad* (see Figure 2-9).

Both of these antennas concentrate your signal (transmit *or* receive) in specific directions. They accomplish this by bouncing the energy back and forth between various *elements*. In a Yagi, the elements are pieces of metal attached to a long boom. Quad elements are wires cut to square shapes and supported away from the boom. The more elements a beam antenna has, the more it concentrates your signal (and the longer it is, too).

Signal concentration in a beam antenna can make up for the lack of power

Indoor Antennas

Antennas always work best when they're exposed to the great outdoors, but many of us are not lucky enough to have a convenient tree, chimney, balcony railing or other outside support. You might even live in a domicile where outdoor antennas are *persona non grata*! What's an eager ham to do?

There's no reason why you can't install your antenna *indoors*. Does your house or apartment have an attic? Attics are often great locations for antennas. The attic doesn't have to be large or fully finished. I managed to squeeze a 2-meter quad antenna and a rotator into an apartment attic that was only five-feet tall at the highest

point. It worked great! Omnidirectional antennas work just as well in attic installations.

If you don't have an attic, consider an unused corner of a spare bedroom, or an empty space in any other room. For instance, you can hang a simple 2-meter *J*-pole antenna (available from several manufacturers) from a curtain rod. It won't be the best antenna you've ever used, but you'll notice a definite improvement. Nearby metal objects, including house wiring and plumbing, may *detune* indoor antennas. This simply means that you may have to fiddle with the tuning assembly on the antenna (if it has one) or adjust the length to obtain an SWR reading below 2:1.

Figure 2-8—A small 2-meter Yagi antenna like this one will focus your signal in the direction of your choice. Note that the antenna elements are perpendicular to the ground. This is called *vertical polarization*. Most FM operators use vertical polarization.

Figure 2-9—This impressive quad antenna will concentrate your power just like a Yagi.

in a transceiver. Let's pretend you're using an H-T with only a 2-W output. A big beam can focus those two measly watts and make it sound as though you're running 200 W!

If beams are so great, why doesn't everybody have one? Well, the focusing ability of a beam is a double-edged sword. A beam antenna that's pointed north, for example, is a poor antenna for signals coming from any other direction. If you want to talk to someone south of your station, you must turn the antenna so it points south. Now we're talking about yet another device and another purchase: An *antenna rotator*.

Rotators are nothing more than fancy electric motors. They're designed to turn heavy loads clockwise or counterclockwise. If cable TV hasn't taken over your neighborhood, you may still see rooftop television antennas equipped with rotators. The rotators are controlled by a selector box that's usually somewhere close to the TV. A three or five-wire cable connects the rotator to the selector box. When you twist the selector knob, the rotator dutifully obeys and turns the antenna to the requested direction.

TV antenna rotators are light-duty devices, but they're usually powerful enough to turn a moderate-sized VHF or UHF beam. For larger beam antennas, you may need to invest in a heavy-duty rotator designed for ham applications. Ham rotators may also come with fancy control electronics complete with digital readouts and so on.

Transmission Lines—The Critical Links

You own a great radio. You've installed a terrific antenna. Now all you have to do is get the RF energy from the radio to the antenna, and vice versa. As straightforward as it sounds, this is the point where many hams go astray.

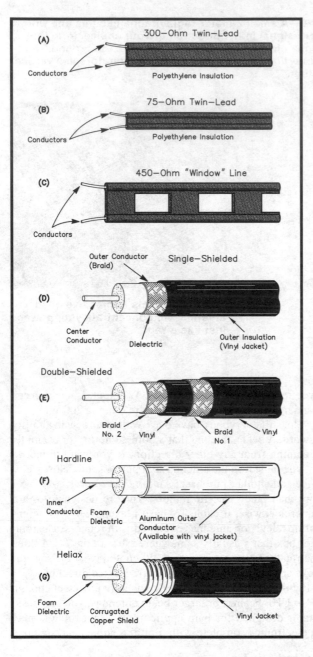

Figure 2-10—Here's the feed-line hall of fame! These are the most common types of lines you'll encounter in amateur use. The twinleads (A and B) as well as the window line (C) are used primarily on the HF bands. You'll use the single (D) and double-shielded (E) lines for most of your VHF and UHF applications. Hardline (F) and Heliax (G) only come into play when you're talking about microwave work, or long feed line lengths at 146 MHz and up.

The *transmission line* is the cable that connects your radio to your antenna. Perhaps you're already familiar with it by another name: *coaxial cable*, or simply *coax*. It's called coax because of the way the cable is constructed. "Coaxial" means "two axis," like two concentric circles. The inner axis is the *center conductor*. This can be a single wire or a tight bundle of stranded wire. The outer axis is the *shield*. The shield can be made of braided wire, or solid, flexible metal. It's separated from the inner axis by plastic or some other insulating material.

"Transmission line" can also refer to cables that are not "coaxial" at all. The flat *ribbon cable* you see on some TV antennas is a type of transmission line. Transmission lines such as these are popular for hamming on the HF bands, but they're not used much for amateur VHF and UHF. Several types of coax and flat transmission lines are shown in Figure 2-10.

I could write an entire chapter on the subject of transmission lines, but you'd want to hang yourself from the nearest shower rod by the time you finished reading it. To tell the truth, transmission line theory is lethally boring. Let's cut to the chase, shall we?

In Table 2-1 I've created a list of coaxial cable recommendations depending on your particular antenna installation. You'll notice that the types of cable I suggest

Table 2-1

Recommended Coaxial Cables

Find your band of interest on the left-hand side of the table, then move right until you reach the length range you need.

Band	Length (feet)				
	<10	10-50	50-100	100-200	>200
50 MHz	RG-58	RG-8/U	8214	9913	9913
146 MHz	RG-58	8214	9913	Hardline	Hardline
222 MHz	RG-58	8214	9913	Hardline	Hardline
440 MHz	8214	9913	9913	Hardline	Hardline
1296 MHz	8214	9913	Hardline	Hardline	Hardline

change according to how much is used and the frequency in question. Even under the best conditions, transmission lines lose a little of the RF energy you put into them. This loss increases as the cable gets longer, the frequency gets higher, or both.

The suggestions shown in the table are based on the assumption that you've installed everything perfectly. The last thing you want to do is increase the losses that could occur in your coax, but there are ample opportunities to do just that!

You see, transmission lines, antennas and radios all have their own *impedance*. In simplest terms, impedance is resistance to the flow of an ac signal. RF energy is an ac signal.

If the impedances of the radio, transmission line and the antenna are the same, you're on easy street. In the case of VHF and UHF ham equipment, that magic impedance number is usually 50 Ω (ohms). But if the impedance somewhere in the system is anything other than 50 Ω, you have a condition called a *mismatch*. A mismatch isn't necessarily a crisis. It all depends on how bad it is.

Any number of things can cause a mismatch: Poorly soldered coax connectors, a mistuned antenna, damaged cable, etc. What we care about most is the effect the mismatch has on the energy in the coax. A mismatch causes a portion of the energy to *reflect* from the location of the mismatch and go bouncing back and forth along the coax. This creates a *standing wave* condition that you can measure with an SWR (Standing Wave Ratio) meter (see Figure 2-11).

When there is no mismatch at all, the SWR will be 1:1. A slight mismatch might show itself with a reading of 1.2:1,

Figure 2-11—A VHF/UHF SWR meter is an essential tool for any FM-active ham. If a serious mismatch condition exists on your feed line, this meter will tell you right away. You install an SWR meter in the line that runs from your radio to your antenna. If you're using an amplifier, the meter goes in the line that runs from your amp to your antenna. In either case, make sure the meter is close to your operating position so you can see it easily.

BNC Connectors

Standard Clamp

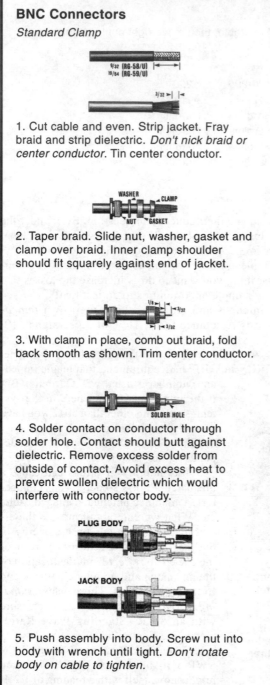

1. Cut cable and even. Strip jacket. Fray braid and strip dielectric. *Don't nick braid or center conductor*. Tin center conductor.

2. Taper braid. Slide nut, washer, gasket and clamp over braid. Inner clamp shoulder should fit squarely against end of jacket.

3. With clamp in place, comb out braid, fold back smooth as shown. Trim center conductor.

4. Solder contact on conductor through solder hole. Contact should butt against dielectric. Remove excess solder from outside of contact. Avoid excess heat to prevent swollen dielectric which would interfere with connector body.

5. Push assembly into body. Screw nut into body with wrench until tight. *Don't rotate body on cable to tighten*.

Improved Clamp

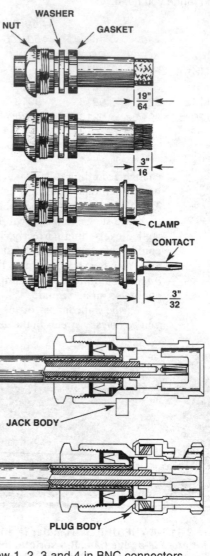

Follow 1, 2, 3 and 4 in BNC connectors (standard clamp) except as noted. Strip cable as shown. Slide gasket on cable *with groove facing clamp*. Slide clamp on cable *with sharp edge facing gasket*. Clamp *should* cut gasket to seal proplerly.

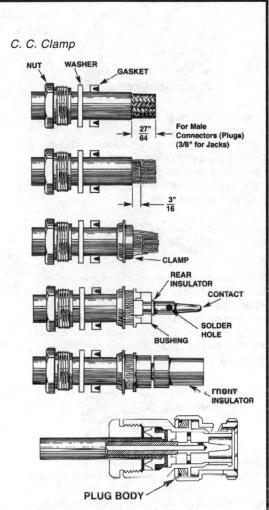

C. C. Clamp

1) Follow steps 1, 2 and 3 as outlined for the standard-clamp BNC connector.

2) Slide on the bushing, rear insulator and contact. The parts must butt securely against each other, as shown.

3) Solder the center conductor to the contact. Remove flux and excess solder.

4) Slide the front insulator over the contact, making sure it butts against the contact shoulder.

5) Insert the prepared cable end into the connector body and tighten the nut. Make sure that the sharp edge of the clamp seats properly in the gasket.

Type N Connectors

Standard Clamp

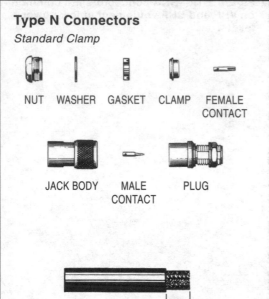

NUT   WASHER   GASKET   CLAMP   FEMALE CONTACT

JACK BODY   MALE CONTACT   PLUG

3) Taper braid as shown. Slide nut, washer and gasket over vinyl jacket. Slide clamp over braid with internal shoulder of clamp flush against end of vinyl jacket. When assembling connectors with gland, be sure knife-edge is toward end of cable and groove in gasket is toward the gland.

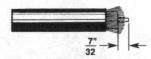

1) Cut cable and even. Remove $^9/_{16}$ inch of vinyl jacket. When using double-shielded cable, remove $^5/_8$ inch.

4) Smooth braid back over clamp and trim. Soft-solder contact to center conductor. Avoid use of excessive heat and solder. See that end of dielectric is clean. Contact must be flush against dielectric. Outside of contact must be free of solder. Female contact is shown; procedure is similar for male contact.

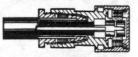

2) Comb out copper braid as shown. Cut off dielectric $^7/_{32}$ inch from end. Tin center conductor.

5) Slide body into place carefully so that contact enters hole in insulator (male contact shown). Face of dielectric must be flush against insulator. Slide completed assembly into body by pushing nut. When nut is in place, tighten with wrenches. In connectors with gland, knife edge should cut gasket in half by tightening sufficiently.

Figure 2-13—Type N connectors are required for medium- and high-power operation on VHF and UHF.

Improved Clamp

Step 1

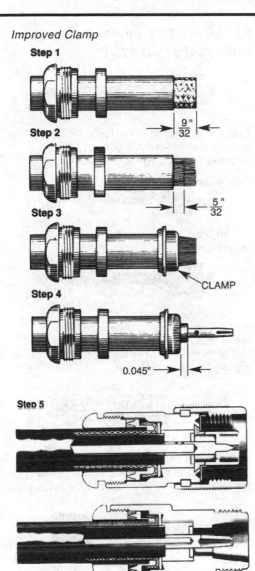

Step 2

$\frac{9}{32}$"

Step 3

$\frac{5}{32}$"

CLAMP

Step 4

0.045"

Step 5

1) Follow instructions 1 through 4 as detailed in the standard clamp (be sure to use the correct dimensions).

2) Slide the body over the prepared cable end. Make sure the sharp edges of the clamp seat properly in the gasket. Tighten the nut.

C. C. Clamp

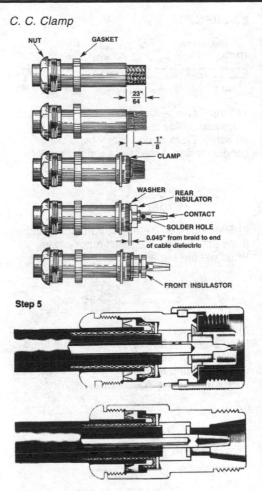

NUT   GASKET

$\frac{23}{64}$"

$\frac{1}{8}$" CLAMP

WASHER REAR INSULATOR

CONTACT

SOLDER HOLE

0.045" from braid to end of cable dielectric

FRONT INSULASTOR

Step 5

1) Follow instructions 1 through 3 as outlined for the standard-clamp Type N connector.

2) Slide on the washer, rear insulator and contact. The parts must butt securely against each other.

3) Solder the center conductor to the contact. Remove flux and excess solder.

4) Slide the front insulator over the contact, making sure it butts against the contact shoulder.

5) Insert the prepared cable end into the connector body and tighten the nut. Make sure the sharp edge of the clamp seats properly in the gasket.

83-58FCP

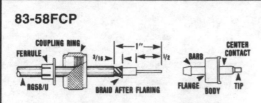

1) Strip Cable—*don't nick braid, dielectric or conductor*. Slide ferrule, then coupling ring on cable. Flare braid slightly by rotating conductor and dielectric in circular motion.

2) Slide body on dielectric, barb going under braid until flange is against outer jacket. Braid will fan out against body flange.

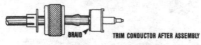

3) Slide nut over body. Grasp cable with hand and push ferrule over barb until braid is captured between ferrule and body flange. Squeeze crimp tip only of center contact with pliers; alternate-solder tip.

83-1SP (PL-259) plug with adapters (UG-176/U or UG-175/U)

1) Cut end of cable even. Remove vinyl jacket 3/4 inch—don't nick braib. Slide coupling ring and adapter on cable.

2) Fan braid slightly and fold back over cable.

3) Position adapter to dimension shown. Press braid down over body of adapter and trim to $3/8$ inch. Bare $5/8$ inch of conductor. Tin exposed center conductor

4) Screw the plug assembly on adapter. Solder braid to shell through solder holes. Solder conductor to contact sleeve.

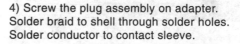

5) Screw coupling ring on plug assembly.

Figure 2-14—You can use crimp-on or solder-on connectors for your RG-58 coax.

which isn't a cause for concern. If you get a reading of 2:1 or higher, however, you need to find out why. Generally speaking, the higher the SWR, the greater the losses in the cable because more of your valuable energy is wasted as it ricochets through the coax. At VHF and UHF frequencies, the loss caused by a mismatch condition can be *huge*! Your radio may also be damaged by a high SWR condition. (Many transceivers are designed to reduce their output if the SWR climbs above 2:1.)

The bottom line: Calculate the length of coax required to connect your radio to your antenna. Figure out the highest band of frequencies where you intend to operate, now or in the future. (Always plan for your *future* needs!) Go to Table 2-1 and select the suggested cable.

Of course, my cable recommendations assume that you'll install your antenna system properly. (You *will*, won't you?) This means attaching the connectors without damaging the coax, setting up the antenna and testing it according to manufacturer's instructions and so on. (See the sidebar, "Antenna Installation Tips.") The result should be an antenna system with an SWR of 2:1 or less.

TURNING UP THE POWER

If your radio is only capable of putting out a couple of watts or so, the day will probably arrive when you realize that you need more RF muscle. Mobile operating at the edge of a repeater's coverage area often makes this need painfully apparent. When you're trying to say something important (to you, at least), it's frustrating to know that your signal is weak into the repeater. You can bet that the folks who are listening to your transmissions are equally miffed!

If you can't rely on your antenna to add some punch to your signal, you'll need to work on the *power* side of the equation. You can compensate for antenna deficiencies by boosting your output to 30 W, 150 W or more. All you need is an amplifier, affectionately known as a "brick."

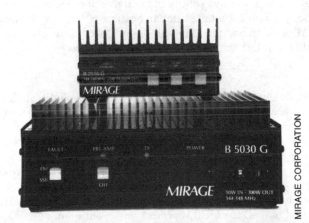

When you need to add some muscle to your signal, consider an amplifier. Two models are shown here. The top unit offers 160-W output for 25-W input. The monster below will boost 50 W to 300 W!

MIRAGE CORPORATION

A Brick is a Brick is a Brick . . .

Bricks are solid-state RF power amplifiers. They're called bricks because they approximate the shape and weight of the venerable baked-mud masonry. Bricks are not terribly complicated. They're basically brute-force amplifiers designed to take your flea power and convert it to tiger power— or at least to a power level equivalent to that generated by a small burrowing mammal.

The average brick features an **ON/OFF** switch that allows you to bypass the amplifier when you're feeling overwhelmingly overconfident about your signal. Some include a receive preamplifier.

Flushing Out Contacts

If you announce that you're "listening" and no one responds, wait and try again a few minutes later. You'll have better luck during the commuting hours in the morning and afternoon. And don't forget about the shyness factor. If no one replies, it's not that they don't like you, they just don't know you.

- Try asking for a signal report rather than

simply stating that you're "listening." A report request gives an otherwise shy ham an extra incentive to call you.

- Join a club that's active in public-service activities. Volunteer for as many events as possible.

- Active contest clubs are also good prospects. Offer your time to assist in several major contests at the club station.

On some bricks you'll find a switch labeled **SSB/FM**. Don't let it confuse you. This ominous selector only affects the amplifier's *hang time*.

When you're operating SSB, your output power changes with the amplitude of your voice. If you're not talking, there's no output. Bricks go into action the instant they sense RF power from your radio. As you babble, the RF power is coming and going at a rapid clip. (Every time you pause to draw a breath, for example, your output drops to nearly nothing.) This causes the hapless brick to jump from transmit to receive like a toad in a hail storm. Relays and other components don't react well to this sort of abuse.

The solution is to make the amplifier remain in the transmit mode for a second or so *after* the output falls to zero. When you flip the switch to **SSB,** the amp waits briefly before returning to the receive mode—just to make sure you don't have more to say.

On FM, your radio is supplying full output for as long as you hold down the microphone push-to-talk button. The amp stays solidly in the transmit mode until you decide to listen once again. So, you don't want the brick to hesitate before it switches to receive. Leave it in the **FM** position and the amp will bounce back to receive the instant you release the mike button. (Use the **FM** selection for packet, too.)

How Much is Enough?

How much output power do you really need? Thirty to 50 W is usually sufficient for mobile work. There are plenty of power amps on the market that will take 5 W or less and kick it up to 30 W. Brand new units in this power class will cost between $60 and $160, depending on how many features you want. Of course, you could run 150 W or more from your car, but I wouldn't recommend it. The intense RF is likely to drive your automotive electronics crazy while it's rearranging your DNA.

For hearth and home the question of power depends on what you want to do. If your goal is to put a solid signal into all the local repeaters with your omni antenna, you can probably remain at the 30-W level. But let's say you enjoy operating FM simplex and you want to expand your coverage. Maybe you've just been elected as the control station for the Klingon Language Net. This is no time to soft-peddle your signal. A jump to the 150-W class may be in order.

YOU'RE ON THE AIR!

To use a repeater, you must know one exists. There are various ways to find these elusive beasts. Local hams can provide information about repeater activity, or you can consult a repeater listing. The ARRL publishes two guides that are indispensable to the active repeater user: *The ARRL Repeater Directory* and the *North American Repeater Atlas*. They're available from your favorite dealer, or directly from ARRL Headquarters. Besides identifying local repeater activity, these books are handy for finding repeaters during vacations and business trips.

Most modern FM transceivers also include scanning features that will let you sweep up and down the band to your heart's content. This is another easy way to find active repeaters. Listen in the early morning and late afternoon.

Remember that you're listening to repeater *output* frequencies. The frequencies you transmit on to use the repeaters—the *input* frequencies—will be different. Many FM radios sold within the last several years handle the differences between input and output frequencies automatically. For example, if you hear a repeater on 147.36 MHz, your radio will automatically select 147.96 MHz as its transmit frequency. The rig "knows" that most repeaters that operate above 147 MHz in the 2-meter band have their input frequencies higher than their outputs. It also knows that the 2-meter input/output frequency separation is 600 kHz. (147.96 MHz −147.36 MHz = .60 [600 kHz])

Naturally, there are exceptions to every rule. You might find a repeater on the 2-meter band that's using something other than a 600-kHz offset. Look up the repeater in *The ARRL Repeater Directory* or the *North American Repeater Atlas* and you'll discover its input frequency. Program the input/output frequency pairs into an available memory channel and you're in business.

Don't Grab the Microphone Yet

Finding a repeater is only half the job. Spend some time really *listening* to it. How busy is it? What sort of conversations do you hear? Do the people sound friendly?

Most importantly, learn the *customs* of the repeater. When someone is fishing for a contact, how do they announce themselves? Is it, "Hey! You stupid fools! I wanna talk to someone!" or something more subtle such as, "WB8IMY, monitoring"? (You'll find that most repeater operators use the words "monitoring" or "listening" as a way to let everyone know that they're in the mood to chat.)

Now, With Trembling Hands . . .

Push must eventually come to shove. The immovable object must meet the unstoppable force. In other words, you must *talk* to someone! You punch in the frequency of the local repeater and listen. Silence. This is the moment of truth. You key the microphone and, in your most confident voice, announce your call sign.

The repeater transmits for a few seconds, then stops. Surely someone is reaching for their microphone. They'll call you in just a few seconds . . . won't they? The seconds stretch into minutes. You announce that you're listening again, this time with added urgency.

Still nothing.

Using *The ARRL Repeater Directory*

If you're traveling with your VHF/UHF rig, the *Directory* is virtually indispensable. How's that for simple? The handy pocket-size book lists thousands of repeaters across the country, from 29.5 through 1200 MHz—voice, packet and beacons.

You'll also find band plans for all VHF/UHF amateur bands, a list of regional and state frequency coordinators and a large list of ARRL Special Service Clubs (many include meeting times and places).

To give you an idea of how the *Directory* is set up, part of page 195 is shown at the right. The listings are for 2-meter repeaters in the state of New York. From left to right, items include (1) location, (2) the repeater's output frequency, (3) the repeater's input frequency (nonstandard splits are listed numerically, standard splits are indicated by plus and minus signs indicating whether the input is 600 kHz above or below the listed output frequency, (4) the call sign of the repeater's trustee, (5) notes about special features and requirements (probably the most important note for casual travelers is the "o," which denotes an "open repeater." If the repeater is "closed," a "c" will be listed. If CTCSS tones are required to access the repeater, they'll usually be listed here, as will a LiTZ symbol for those repeaters using the "long-tone zero" for emergency activation. The complete list of notes appears on pages 19 and 20), (6) the club or individual sponsoring the repeater.

The ARRL Repeater Directory is a trav-

208 144-148 MHz NEW YORK					
Location	Output	Input	Call	Notes	Sponsor
Grafton	147.180	146.160	WS2B	o(ca) elzRA	Rens.RACES
Grafton	147.180	447.200	WS2B	o(ca) elzRA	Rens.RACES
Hoosick Falls	147.345	+	KB2KYZ	orA	Rens.RACES
Poestenkill	145.370	–	N2JXO	orA	Rens.RACES
Troy	146.760	–	KB2HPX	oreA	Rens.RACES
Troy	146.760	147.780	KB2HPX	oreA	Rens.RACES
Troy	146.760	447.200	KB2HPX	oreA	Rens.RACES
BINGHAMTON					
Binghamton	145.210	–	N2LZM	o	BURS
Binghamton	146.730	–	K2TDV	oe	
Binghamton	146.820	–	WA2QEL	oae	SVARA
Binghamton	147.390	+	W2OW	oae	
Maine	147.075	+	KB2KW	ol	S.TIR ARES
Owego	146.760	–	W2VDX	o	TIOGA ARA
Port Crane	145.290	–	WA2QEL	o	BARA
Walton	147.315	+	W2LZ	oer	Walton RA
BRONX					
Bronx	147.240	+	KA2NYR		WA2DUO
Bronx#	146.820	–	N2EQX	o(ca)	+KA2TQJ
Brooklyn	145.230	–	W2CMA		W2CMA
Brooklyn	146.730	–	KW2V	o	KW2V
CAPITOL DISTRICT					
Albany	145.190	–	KM2H	oe	ALB ARA
Albany	146.790	–	WA2BFI	o(ca)el	SEEN
Albany	147.120	+	WB2ZCM	ore	ALB RACES
Broadalbin	145.430	–	WB2BGI	o	
Cohoes	147.150	+	WB2TDG	o	
Delmar	146.640	–	WB2VJB	or	BETH.RACES
Duanesburg	146.865	–	WA2AAU	o	NOAHS ARC
Gloversville	146.700	–	WA2ZJF	oar	TRYON ARC
Grafton	145.310	–	K2CBA	o	GURU
Grafton	147.375	+	K2CBA	ol	GURU
Knox	147.315	+	WA2WNI	o	GRAFTONUR
Lake Nancy	147.360	+	KC2IV	o	
Loudonville	146.760	–	N2CHH	orl	RENSRACES
Schenectady	147.060	+	K2AE	oaer	SARA
Schenectady	147.300	+	WA2AFD	oa	
Troy	145.170	–	NY2U	or	TROY ARA
Troy	145.250	–	WB2BQW	oaez100.0	
Troy	145.330	–	KA2NMP	olr	NiMo ARC
Troy	146.820	–	W2SZ	oe(ca)	RPI RC
Troy	146.940	–	W2LWX	oael	TEL PIONRS
West Grafton	146.835	–	WA2WNI	o	GURU
CATSKILLS					
Hunter	145.150	–	WB2UYR	oer	MTNTOPARA
Jewett	145.450	–	KA2PAP	o	
Roxbury	146.985	–	K2AGF	oe	Mrgrtvl RC
CENTRAL					
Canastota	146.670	–	N2JEU	oaeyzx	
Cazenovia	147.075	+	N2LZI	oe151.4	N2LZI
N Columbia	145.110	–	WB2KMH	oaer	Herk ARES

eling ham's "best buy." If you travel a lot, you'll wear your copy down to a frazzled nub! Get one from your local dealer or directly from ARRL HQ.

Again the lonely minutes pass. Maybe you just picked a bad time. You'll try again in an hour or so. As you reach for the **POWER** switch, the repeater suddenly comes to life.

"WB8ISZ this is WB8SVN. You around, Dave?"

"WB8SVN from WB8ISZ. I'm here. Did you just get off work?"

Now you feel a new emotion—anger! It's a safe bet that one of these two guys were listening before. Why didn't they answer you? Is it because you're a new ham?

The Shy Communicators

Hams pride themselves on their ability to communicate, yet there is an odd contradiction: many hams are painfully shy! If you don't believe this, go to any hamfest. Chances are, you'll see hams whose call signs you recognize—hams who are constantly chattering on the local repeaters. So why are these same hams wandering around so quietly? When you approach them, why do they seem so ill at ease and reluctant to talk?

The answer lies in the nature of Amateur Radio itself. With the exception of visual modes such as ATV, no one can see you when you're on the air. You could be holding a conversation with someone while wearing little more than your underwear. They'd never know! In other words, ham radio allows us to hold the world at arm's length while still maintaining contact. It acts as a filter and a shield for those who are uncomfortable with close, personal communications.

Breaking through the shyness barrier to communicate with a stranger is difficult. Think back to your school days. When the teacher asked for student volunteers for a project, why did you hesitate? Perhaps you wanted to see if anyone else was willing to join you. No one wants to be the first to raise their hand!

A similar situation occurs on repeaters. When you announced that you were listening, a dozen people may have heard you. No one recognized your call sign, though. You're a stranger, an unknown. It's as though the teacher just got on the repeater and asked for volunteers to speak to you. Who will be the first to step forward?

Where's the Action?

When you're looking for repeaters, set your radio to scan between the following frequencies . . .

6 Meters:

    51.62—51.98 MHz

    52.5—52.98 MHz

    53.5—53.98 MHz

2 Meters:

    145.20—145.50 MHz

    146.61—147.39 MHz

222 MHz:

    223.85—224.98 MHz

440 MHz:

    442.00—445.00 MHz

    447.00—450.00 MHz

1296 MHz:

    1282—1288 MHz

For FM simplex activity, try these national calling frequencies:

    52.525 MHz, 146.52 MHz, 223.5 MHz, 446 MHz, 1294.5 MHz

For many hams, the familiar line of reasoning is, "Hmmm . . . I don't know this guy. What would I say to him? Nah . . . I'll wait. I'm sure someone else will give him a call." The problem is, when all the hams on the repeater feel this way, no one replies!

And so it goes on repeaters throughout the country. The problem isn't you *per se*, it's that fact that you're a stranger. So how do you make the transition from stranger to friend?

Breaking the Ice

If you keep announcing that you're "listening," someone is bound to come back to you eventually. This could take a long time—especially if you're trying to start a conversation during less popular hours. To really break the ice and shed your "stranger" label, you need to assert yourself on the air. That is, you need to become part of an existing conversation.

Listen to the repeater during the early morning and late afternoon. That's when it's likely to be used the most. As you hear stations talking to each other, listen for an opportunity to contribute something—even if it's just a question. Let's say that you find two hams discussing computers . . .

"KR1S from WR1B. Well, I'm definitely going to pick up some extra memory at the show tomorrow. I figure I need at least 2 megabytes."

"I don't know, Larry. I think 4 megabytes would be a better choice for the kind of software you're running."

Even if you don't own a computer, I bet you can think of a question that will give you an excuse to join the conversation. In the pauses between their transmissions, announce your call sign.

"WB8IMY"

"Well, there's a new voice. Ah . . . WB8IMY . . . I think it was . . . this is KR1S. How can I help you?"

"Hello. My name is Steve and I live in Wallingford. I'm thinking about buying a computer for my Amateur Radio station, but I'm a little confused. You guys seem knowledgeable. Can you give me a recommendation?"

Perfect! Stroking a person's ego is the best way to get them talking. With luck, these fellows will be more than happy to show off their expertise. Just keep the questions and comments coming.

If you engage in enough of these conversations on the same repeater, you'll gradually melt through the shyness barrier. In time, your call sign will be as familiar as any other. When you say, "WB8IMY listening," you'll have a much better chance of getting a response. After all, they'll *know* you.

Getting Involved

Another way to establish yourself is to become involved in club activities. Look for a local club that's active in public-service events. Attend the meetings regularly and be prepared to volunteer whenever they ask for help.

Don't worry about your lack of experience in public-service operating. Believe me, it isn't that difficult. You'll be told exactly what to do and, in most cases, an experienced ham will be nearby.

My first public-service activity was a canoe race in my home town of Dayton,

Ohio. I was the new face in the club and I was new to ham radio. When they asked for volunteers, it took a great deal of courage to raise my hand. Boy, am I glad I did!

The race organizers needed "checkers" at various points along the river. It was our task to make sure that each canoe passed our checkpoint safely. I was stationed with my FM transceiver at an isolated rural bridge over the Miami River. As each canoe passed beneath me, I checked it off my list and relayed the information to the net-control station. The sun was shining, a gentle breeze was blowing through the trees and I felt terrific! Here I was, an Amateur Radio operator, doing an important job with my fellow team members.

After the race, we all met at a local pizza restaurant and swapped stories. Someone asked if I wanted to be part of the communications team for the March of Dimes walk-a-thon the following weekend. Why not? After participating in several public-service events, everyone knew me by name and call. There was never a shortage of someone to talk to on the repeater.

Acknowledging Stations

If you're in the midst of a conversation and a station transmits its call sign between transmissions, the next station in queue should acknowledge that station and

The Silent Service

Subaudible means "below audible"; below the range of human hearing. Repeaters use these low-frequency audio tones for special purposes. Your ears can't hear them, but a repeater has no problem detecting their presence. Subaudible tone generators and decoders are often lumped together under the term "continuous tone-coded squelch system," or CTCSS for short. Modern FM transceivers include CTCSS generators (encoders), or at least provide them as options.

Most repeaters use subaudible tones as an effective way to deal with interference. When a repeater is using a CTCSS detection system, it will only repeat signals that carry the proper subaudible tone. An interfering signal, such as *intermod* caused by a nearby commercial transmitter, will be ignored. So, if you can't seem to use a particular repeater—and you're sure you are within its range—it might be using a CTCSS system. You'll need to find the correct subaudible tone and program it into your transceiver. Some radios will scan for CTCSS tones on a repeater's signal and display the tone frequency. If your radio lacks this neat feature, check for tone information in *The ARRL Repeater Directory*, or ask around at your next club meeting.

The W1KKF repeater in Wallingford, Connecticut, uses subaudible tones in an interesting way. The problem centers on a powerful repeater in the New York City area that shares the same frequencies with W1KKF. Although the users of the New York repeater don't usually key up W1KKF, their repeater is powerful enough to be heard throughout a large part of W1KKF's coverage area when the band is open. As you can guess, anyone with a sufficiently sensitive receiver is driven insane by signals from two repeaters at once!

The solution? W1KKF transmits a 162.2-Hz subaudible tone whenever it repeats a signal. Hams who own radios with CTCSS squelches can set their rigs to respond only to signals that carry the 162.2-Hz tone. In other words, they won't hear a peep unless the signal is coming from the W1KKF repeater.

permit the newcomer to make a call, or join the conversation. It's discourteous not to acknowledge him and it's impolite to acknowledge him but not let him speak. You never know; the calling station may need to use the repeater immediately. He may have an emergency on his hands, so let him make a transmission promptly.

The Pause That Refreshes

A brief pause before you begin each transmission allows other stations to participate in the conversation. Don't key your microphone as soon as someone else releases his. If your exchanges are too quick, you'll block other stations from getting in.

The "courtesy beepers" on some repeaters compel users to leave spaces between transmissions. The beep sounds a second or two after each transmission to permit new stations to transmit their call signs in the intervening time period. The conversation may continue *only after the beep sounds*. If a station is too quick and begins transmitting before the beep, the repeater may respond to the violation by shutting down! So, if the repeater uses a courtesy beep, wait until you hear it before you continue talking.

Brevity is the Soul of Wit

Keep each transmission as short as possible. Short transmissions permit more people to use the repeater. All repeaters promote this practice by having timers that "time-out," temporarily shutting down the repeater if someone babbles beyond the preset time limit. With this in the back of their minds, most users keep their transmissions brief!

Learn the length of the repeater's timer and stay well within its limits. The length may vary with each repeater; some are as short as 15 seconds and others are as long as three minutes. Some repeaters automatically vary their timer length depending on the amount of activity on the system; the more activity, the shorter the timer.

Because of the nature of FM radio, if more than one signal is on the same frequency at one time, it creates a muffled buzz or an unnerving squawk. If two hams try to talk on a repeater at once, the resulting noise is known as a "double." If you're in a roundtable conversation, it is easy to lose track of which station is next in line to talk. There's one simple solution to eradicate this problem forever: *Always pass off to another ham by name or call sign*. Saying, "What do you think, Jennifer?" or "Go ahead, 'YUA" eliminates confusion and avoids doubling. Try to hand off to whoever is next in the queue, although picking out anyone in the roundtable is better than just tossing the repeater up for grabs and inviting chaos.

Autopatching

Autopatches allow mobile and portable operators to place telephone calls through the repeater. Let's say I'm cruising around town and I see a car accident. By pressing three keys in sequence on my H-T's keypad, I can activate the repeater autopatch. I'll hear the dial tone when I release the push-to-talk (PTT) switch. Now I can transmit again, this time using the keypad to dial 911.

Although they may seem similar, a repeater autopatch is not the same as a cellular telephone. They both use RF, but the similarity ends there. Autopatches are comparable to old-fashioned "party lines." When you make an autopatch call, *everyone* gets an earful of your conversation. Cell phones are relatively private by comparison. Autopatches are only *half-duplex* devices. This means that you and the person you've called must take turns talking. While you're babbling away, your buddy can't get a word in whatsoever. Cell phones, on the other hand, are *full duplex;* you can interrupt each other at will. Finally, autopatches are intended for short-term use. You make your call, speak your piece and get off. You can chat all day on a cell phone—if you can recover from the shock when you see your bill.

Just because you're able to use a repeater, don't expect free access to its autopatch. Most repeater groups require autopatch users to be paid members. As a member in good standing, you'll receive the "secret" codes that operate the autopatch. Let's say that you've just paid your dues and you've been told that the repeater autopatches codes are *-9-1 to turn it on and #-3-6 to turn it off. Here's how an autopatch conversation might sound . . .

"WB8IMY to access the 'patch"

(I announce my intention and listen for a second, just to make sure that no one needs to use the repeater for an emergency.)

*-9-1

(While holding down the microphone button, I press the access code on the radio keypad. I release the mike button and I hear the repeater sending a telephone dial tone.)

5-5-5-6-9-3-8

(I hold down the mike button again and punch in the telephone number. As I release the mike button, I hear the telephone ringing.)

"Hello?" my friend answers.

"Hi, Alan. This is Steve. I'm calling you by ham radio from my car. We have to take turns talking, okay? I can't hear you when I'm speaking."

"I understand."

"I'm about 15 minutes from your house. Mind if I stop by and pick up that old mobile antenna?"

"That's fine, I'll have it ready. Can you stay for lunch?"

"I'd be glad to. See you soon."

"Bye, Steve"

#-3-6 (I send the code that shuts down the autopatch.)

"WB8IMY clearing the 'patch." (Just another courtesy to let everyone know that I'm finished.)

Free Speech?

Can you use an autopatch for *any* purpose? Well, consider that you're actually using the facilities of someone else's Amateur Radio station. That's exactly what a repeater is, remember? The control operator has the last word about what takes place on his or her repeater—including the autopatch.

This is another instance where it pays to learn the customs of your repeater. For example, some control operators don't mind if you use the autopatch to order a pizza. Others mind very much! Listen for a while and learn the pattern. If in doubt, don't do it until you can speak to the control operator.

Beyond the preferences of the control operator is the mighty arm of the FCC. When it comes to autopatches, the FCC has set up several legal taboos.

1. Thou shalt not use an autopatch in the commission of a crime. ("Hey, Charlie! I'm just a couple of blocks away. Can I pick up that kilo of coke you promised me?")

2. Thou shalt not use an autopatch to avoid long-distance telephone charges. ("It's long distance, but don't worry. I'm using a ham autopatch.")

3. Thou shalt not use an autopatch for business purposes. ("This is Gwen. Has our shipment of banana leaves arrived on the dock yet? Don't forget to check the bill of lading before you accept it.")

Simplex Patches

Simplex autopatches use a single frequency to provide a link between mobile or portable transceivers and local telephone systems. Unlike repeater autopatches that operate in full duplex applications (transmitting and receiving simultaneously on separate frequencies), a simplex autopatch uses the same frequency for both transmission *and* reception.

Many simplex autopatches accomplish this feat by cycling from transmit to receive in fractions of a second. During each brief receive cycle, the patch checks the radio for the presence of a carrier. If a carrier is detected, it assumes that the operator of the mobile or portable station is talking and it feeds this audio to the telephone line. If not, it continues to transmit the telephone audio while monitoring the receiver for activity. (Simplex autopatches include timers that will shut down the system if nothing is heard from the receiver after a certain amount of time.)

Assuming that the transceiver can cope with such rapid switching, it's possible to carry on a conversation with minimal interference—although each party is likely to hear pulsing "clicks" in the audio. Both parties must also take turns talking. Like a repeater autopatch, a simplex autopatch is controlled through the use of DTMF (*TouchTone*) tones sent from your transceiver keypad.

Simplex autopatches are often used in situations where hams want telephone access without the cost and hassle of building a complete repeater system. A simplex autopatch requires only a transceiver, an antenna and the means to connect to the telephone line. They don't provide the smooth, reliable operation of a repeater autopatch, but they're better than having no autopatch facilities at all.

With declining prices, simplex autopatches are growing in popularity. The problem, however, is finding available frequencies for them. Operating a simplex autopatch amounts to remote control of a station, so you must use them at 222.15 MHz or above. Setting up an autopatch system on whatever frequency strikes your fancy is a formula for trouble.

Reverse Autopatches

A *reverse autopatch* is exactly what the name implies. It accepts incoming calls from the telephone system and then sends a signal over the air (through the repeater) to indicate that someone has called. By responding with the proper tones from their radios, any hams who are monitoring the repeater can access the autopatch and answer the call.

The user of a reverse autopatch is essentially controlling the repeater remotely (via the telephone lines). If the caller is a ham, this doesn't present a problem—assuming that all other legal requirements have been met. But if the caller is *not* a ham, you suddenly have a situation where a nonamateur is in control of a ham station without a control operator present. The FCC expressly forbids this!

Some repeaters get around this problem by using a kind of voice mail system. A nonham can call the repeater and leave a message. No signal is sent over the air. It's up to the hams to periodically access the repeater and check their voice mail to see if any messages are waiting.

A Packet For You

Computers in the future may weigh no more than 1.5 tons.

—*Popular Mechanics*, forecasting the relentless march of science, 1949

If you're into computers at all, you'll love this chapter. If not, skip ahead to Chapter 4 where you'll learn about all the fun, weird stuff you can do on VHF and UHF . . . *without FM or repeaters*.

Still with me? Good. Cuddle up to your PC, Mac or whatever, and I'll tell you a story about computers, radios and communicating without telephone lines.

SWAPPING INFORMATION

If you had a file that you wanted to transfer to a friend's computer several miles away, how would you do it? I bet you'd make a copy on a diskette and deliver (or send) it to your friend, right? Fair enough.

But what if you were too lazy to drive over to your friend's house, or too cheap to spring for a few postage stamps? (I don't mean to imply that you have a dysfunctional personality. This is just an analogy!) You could hook up a *modem* to your computer and, assuming your buddy has a modem as well, send the file to him over the telephone line.

Modems simply take data from your computer—in the form of shifting voltages—and transform it into shifting audio tones. Once you've made this magical transmutation, it's easy to send the information over the telephone. After all, if a telephone line can send your voice, it can send just about any other type of audio signal. Depending on the speed your modem is running, the tones sound like a chorus of buzz saws or hissing tomcats. The telephone lines handle these raucous tones just fine.

The same modems also receive tones from telephone lines and translate them back to shifting voltages. When fed to your computer, these shifting voltages are interpreted as data. There you have it! Two computers with modems can communicate just about anything between each other . . . as long as a telephone line is available.

Snip! Snip!

Oops! There goes that telephone line! Too bad. How will you get the data to your buddy now?

Well, you're a ham and he's a ham. You have those fancy VHF FM transceivers sitting on your desks. Can you use a radio link to replace the telephone line? You bet!

The first thing you must do is replace your modems with *terminal node controllers*, otherwise known as *TNC*s. These devices are quite similar to telephone modems, but they're designed to transfer data over radio links. TNCs connect to your computer (via a serial port) and to your FM transceiver (see Figure 3-1).

The primary function of a TNC is to receive data from your computer (whatever you're typing on the keyboard, for example) and process it into something that can be sent over a radio link. Conversely, a TNC translates the audio signal from your radio and creates digital information your computer will understand. A TNC also handles the flow of information over the link.

Because it contains its own software (called *firmware*) and a microprocessor, a TNC requires little from your computer. In fact, all you need is a *terminal* program to enable your computer to "talk" to the TNC. There are plenty of terminal programs available. You can even use a terminal program that's designed for telephone modems.

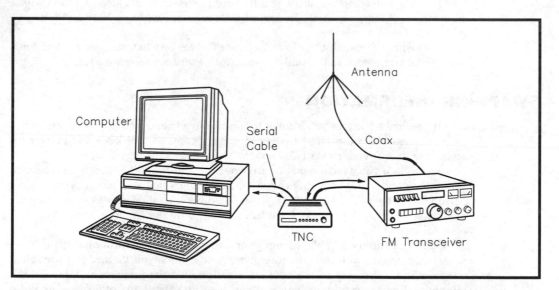

Figure 3-1—The simplest packet station is composed of three items: a computer, a TNC and an FM transceiver. The TNC is a kind of radio modem, taking data from the computer and translating it into audio tones for the transceiver. The TNC also takes audio from the transceiver and converts it into data for the computer.

But What About Packet?

What, you may ask, does all this TNC detail have to do with packet? What is a "packet" anyway?

Let's say that I wanted to transmit the contents of this book from my computer to your computer. I could establish a radio link to your station and, using a spectacularly dumb radio modem, send everything to your computer in one transmission. It sounds easy so far, doesn't it?

Bonnie King, KD4BNV, keeps in touch with friends through packet radio. (*AC4HF photo*)

Why don't we make our example a little more challenging? We'll add two more hams on the same frequency. These guys don't want to send books to each other, but they do want to swap lengthy opinions on the latest music. I'm quick on the trigger, however, so my station transmits first. The entire text of this book is suddenly flying across the airwaves in a long, continuous stream of data. My signal is occupying the frequency, so the other hams have to stand by until I'm finished.

At long last my transmission ends. Now the other hams can send their information. But wait! During my transmission there were bursts of noise and interference that destroyed some of the data. You can't have a book with errors, so there is only one thing I can do: send the entire book again! Without hesitation, my computer switches on the transmitter and repeats the outrageously long transmission.

By this time everyone is staring at their computer screens and muttering, "This is ridiculous. There must be a better way." Of course, they're right!

Let's try again, but this time we'll use *packet* techniques (finally!) and a TNC to get the job done. The first step is to establish a connection between our stations. We say our stations are *connected* when the TNCs at each end "talk" to one another and establish the transmission and reception *protocols*. (A protocol is a formal, standardized way of doing something. Clubs often use Robert's Rules of Order to keep meetings civilized. Robert's Rules of Order are a set of protocols.) In the case of packet, the protocol is known as *AX.25*.

My TNC begins by digesting the text of the book a little at a time and constructing a number of "byte-sized" (pardon the pun) packets. These packets are little more than chunks of data assembled in a manner that complies with the AX.25 protocol. They're lined up in proper order like airplanes sitting on a taxiway waiting to take off. Within a fraction of a second, the first packet is on its way.

Your TNC receives the packet and checks it for errors. In the meantime, my TNC has started a timer that's counting down to zero. If I don't hear from you before the timer reaches zero, the packet is sent again. For the moment we'll assume that my first packet arrived error-free. Your station communicates this fact by sending a special *acknowledgment packet*. When I receive your acknowledgment, my next packet is transmitted.

"Hold on!" you say. "What happens if your station doesn't hear my acknowledgment for some reason?" If your acknowledgment arrives distorted and unreadable—or not at all—the TNC timer simply continues to count down to zero. Eventually, my station will resend the first packet. When you receive the first packet again, your station will, in effect, say, "What's going on here? I already have this packet. Send the next one!" It indicates its displeasure by transmitting what is known as a *reject* packet. When I receive the reject, my station will automatically send the next packet.

Through this process of timers, acknowledgments and rejects, all of my packets will arrive at your station in one piece—and vice versa. When signals are strong and clear at both stations, the packets are transmitted and acknowledged in rapid-fire sequence.

During the silent periods between our transmissions, the other hams can send their packets, too. Packet is extremely patient. If a station has a packet to send and another station is transmitting on the frequency, it will wait until the transmission stops before trying to send its data. Packet is also very persistent. A packet will be transmitted and retransmitted many times before the system finally gives up and breaks the connection.

From Little Acorns, Mighty Oaks . . .

When something exciting is taking place, everyone wants to jump on the wagon. It starts with a couple of packet radio fanatics sending messages just as I've described. They're joined by even more packet aficionados. Before long, someone gets around to setting up a bulletin board system (BBS) where everyone can exchange information and work other interesting magic. Not content with limited coverage, other packeteers set up relay stations known as *nodes*. What was once bound is now unbound! A friendly exchange of data has become a full-fledged packet network linking hams throughout the city, county or state!

This is where we stand today. In less than two decades packet has moved from an experimental mode to one of the most popular means of amateur communication. Thanks to the help of the Internet, some packet networks have gone nationwide and worldwide.

SO WHAT CAN I DO WITH PACKET?

The answer depends on your interests! Every amateur has different uses for packet. Some view packet as a tool to enhance their overall enjoyment of the hobby. Others take it further, using packet as their primary means of communication.

To make things simple, let's list the most popular applications of packet as it exists today:

- Enjoying live keyboard-to-keyboard QSOs.
- Accessing bulletin boards to exchange messages with other amateurs and read general-interest bulletins. Some packet bulletin boards offer additional services such as electronic call-sign directories and magazine bibliographies.
- Using DX *PacketClusters* to hunt HF or VHF DX. Through *PacketClusters* you can determine which DX stations are on the air at the moment—and where they are! DX *PacketClusters* support other useful features as well.
- Tapping elaborate ham and nonham (Internet) networks to exchange files and other information.

Before you can do any of these marvelous things, however, you need to have the proper equipment.

BUILDING YOUR PACKET STATION

A basic packet station consists of an FM transceiver, a computer and a terminal node controller (TNC). For most packet operating, fancy antennas are *not* required. In Chapter 2 we discussed the various types of antennas used for FM voice. You can use these same antennas for packet radio. In most cases, an omnidirectional antenna (one that radiates your signal in all directions, remember?) is a good choice. You only need to consider a beam antenna if you live a substantial distance from a network node or a BBS.

The Computer

Although most amateur packeteers rely on PCs or Macs, just about any computer will do the job if it meets the following requirements:

- It must have a *serial port*. This is the pathway the computer will use to communicate with the TNC and vice versa.
- It must be able to run *terminal* software. A terminal is a simple keyboard and monitor package that's used to communicate with computer systems. Since your TNC contains a microprocessor computer, it also requires a terminal to communicate with its human operator—you! By running a terminal program, your computer will behave just like a terminal as far as the TNC is concerned.

Table 3-1 lists packet software packages for a variety of computers. These are designed with amateur packet in mind. However, if you already own software for communicating with telephone BBSs, chances are the software will also be usable for packet. For example, if you're running Microsoft *Windows*, the terminal program included with the system is fine for talking with a TNC.

If you have a computer gathering dust in your closet, drag it out and put it on packet! You can also find Commodores, Tandy Color Computers, Apples and old IBM PCs selling for bargain prices at hamfest flea markets.

The Radio

In the beginning, amateur packet radio moved data at a speed of 1200 baud. That was considered fast in the early '80s, even by telephone-modem standards. But as telephone modems became faster (now up to 28,800 baud and beyond), amateur packet remained stuck at 1200 baud.

The problem is centered in the radio. Almost any FM transceiver will serve as a packet radio if you're operating at 1200 baud. The only exceptions are rigs that can't switch from transmit to receive quickly, or rigs that distort the packet tones during transmission or reception. Fortunately, these problems are not common. You can even use hand-held FM transceivers for packet if you connect them to outdoor antennas.

To set up an FM radio for 1200-baud packet, you simply feed the transmit audio from the TNC to the microphone jack along with the PTT (push to talk) keying lines. Receive audio is taken from the external speaker jack and routed to the TNC (see Figure 3-2).

Table 3-1

Packet Terminal Software

You don't need *packet* terminal software to communicate with your TNC. Any terminal software package will do the job. Even so, software that's written for amateur packet radio will make your experience that much more enjoyable.

Many TNC manufacturers supply or sell software for their products. If you do not wish to use their software, or if their software won't run on your computer, consider the programs shown below.

Please note that the products and addresses shown below are subject to change. Contact the distributors to verify availability before ordering.

Apple II

APR: Send a blank, formatted diskette and a postage-paid, self-addressed disk mailer to Larry East, W1HUE, PO Box 1355 Rimline Dr, Idaho Falls, ID 83401.

Apple Macintosh

MacRATT: Advanced Electronic Applications (AEA), PO Box C-2160, Lynnwood, WA 98036-0918
Virtuoso: James E. Van Peursem, KEØPH, RR #2, Box 23, Orange City, IA 51041

Atari

PK2: Electrosoft, 3413 N Duffield Ave, Loveland, CO 80538

Commodore

TNC64: Texas Packet Radio Society, PO Box 50238, Denton, TX 76206-0238
DIGICOM>64: A & A Engineering, 2521 West LaPalma, Suite K, Anaheim, CA 92801, tel 714-952-2114
MFJ Enterprises, Box 494, Mississippi State, MS 39762 tel 601-323-5869
(*DIGICOM>64* is a TNC emulation program that causes the computer to behave as a TNC. The only external device required is an inexpensive modem interface. The software and modem can be purchased separately.)

IBM PC

ComTreK for *Windows*, PO Box 4101, Concord, NH 03302. $29.95.
LAN-LINK: Joe Kasser, G3ZCZ, PO Box 3419, Silver Spring, MD 20918. Evaluation copy available for $5.
PK GOLD Enhanced: InterFlex Systems Design Corp, PO Box 6418, Laguna Niguel, CA 92607-6418.
PC-Pakratt II: Advanced Electronic Applications (AEA), PO Box C-2160, Lynnwood, WA 98036-0918.
BayCom: PacComm Inc, 4413 North Hesperides St, Tampa, FL 33614-7618, tel 813-874-2980
(BayCom is a TNC emulation program.)

Tandy Color Computer

COCOPACT: For more information, send a self-addressed, stamped envelope to Monty Haley, WJ5W, Rte 1, Box 150-A, Evening Shade, AR 72532

Many hams have been trying to push the packet throttle to 9600 baud, but packeteering at this speed is a different kettle of fish. You *cannot* feed a 9600-baud data signal from a TNC to the microphone jack of an FM transceiver. The circuitry in the microphone amplifier section trashes the signal and makes it worthless on the receiving end. Instead, you must bypass the microphone amps and inject the signal directly at the FM *modulator* stage.

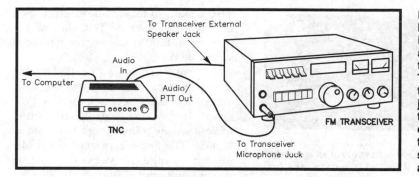

Figure 3-2—For 1200-baud packet operating, the audio output and PTT (push-to-talk) cables from the TNC connect directly to the microphone jack of any FM transceiver. Another cable from the transceiver's external speaker jack supplies audio to the TNC.

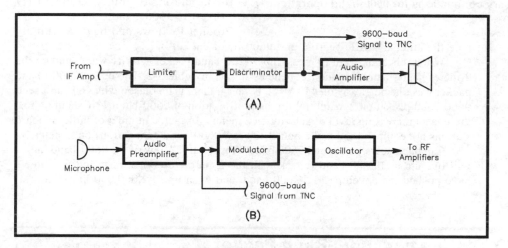

Figure 3-3—Connections for 9600-baud packet are very different from those used for 1200-baud activity. Receive audio for the TNC must be tapped at the output of the transceiver's FM discriminator. Audio *from* the TNC must be injected at the input of the transceiver's modulator stage. Many "9600-ready" radios include data jacks that make these connections for you.

Receiving a 9600-baud signal is just as tricky. You can't tap the audio at the speaker jack like you can with a 1200-baud signal. The receiver's audio stages distort the signal and render it as little more than an annoying hiss in your speaker. You must grab the 9600-baud signal right at the FM *detector,* commonly known as the *discriminator* (see Figure 3-3).

For years the only way to jump to 9600 baud was to modify FM transceivers as I've just described. But the idea of performing brain surgery on a multihundred-dollar radio scared off many a curious packeteer. As a result, 9600-baud packet was a haven for the brave elite while everyone else plodded along at 1200 baud.

The times they are a-changin', thank goodness! Several ham manufacturers are now selling FM transceivers that are designed to be compatible with 9600-baud packet

The Azden PCS-9600D FM transceiver is a typical 9600-baud ready radio. It includes a data jack (on the front panel, in this case) that makes the necessary connections for 9600-baud operation.

TNCs. Many of these radios feature special *data jacks* for your TNC. Just wire a plug according to the instructions and you're on the air!

Beware that some so-called "9600 ready" radios may lack the performance necessary to operate well at this speed. In fact, many of the first crop of 9600-ready transceivers were rather dismal! The manufacturers have made design changes, however, and the newer 9600-ready radios really live up to the claim. If you're going to try packet at 9600 baud, check the *QST* Product Reviews and buy accordingly (see the sidebar "BER Testing and 9600-baud Packet").

When this book went to press, most 1200-baud packet activity was found on the 2-meter band, mainly on the frequencies shown in Table 3-2. The bulk of 9600-baud packet was blasting away on 440 MHz. Even so, there is no reason why you can't send 9600-baud packet on 2 meters. With the proliferation of 9600-baud TNCs and radios, there is an increasing level of activity on 2 meters. In fact, a number of bulletin board stations are equipped to handle both 1200 *and* 9600-baud connections on 2 meters.

Can packeteers go beyond 9600 baud? Absolutely! A few of the 9600-baud radios will operate at 19,200 baud (check with the manufacturers). Pioneering packeteers have pushed the envelope to 56,000 baud, and even up to 2,000,000 baud (no kid-

BER Testing and 9600-Baud Packet

If an advertisement says that a radio is "9600-baud ready," does this mean that it's necessarily a good performer? When it comes to overall performance, you can't cut corners at 9600 baud. The transmitter must be able to send the 9600-baud signal without distortion. The receiver must also be able to receive the 9600-baud signal without including distortion of its own. When you add distortion to packet signals, you begin to see errors in the bits of information being sent or received. We measure the effects of the distortion in terms of the *bit error rate*, or *BER*. The higher the BER, the poorer the performance.

Beginning with the May 1995 issue, all *QST* magazine Product Reviews include the results of BER testing for transceivers that claim to be 9600-baud ready. (*QST* is the only Amateur Radio magazine that publishes BER test results.) BER testing is fairly complicated and so are the results. To make it easier for you to interpret the numbers, here are two rules of thumb:

For receiver performance, look for a "BER @ 16-dB SINAD" of 1×10^{-5} or less.

For transmitter performance, look for a "BER @ 12-dB SINAD + 30 dB" of 1×10^{-5} or less.

Watch out for those exponential numbers! A BER of 1×10^{-3} is *worse* than a BER of 1×10^{-5}! Add the appropriate number of zeros and you'll see what I mean. Would you rather own a transmitter with a BER of less than 1 bit for every 1000 or 1 bit for every *100,000*? One faulty bit for every 100,000 bits (1×10^{-5}) looks pretty good to me!

Table 3-2
Packet Hot Spots!

6 meters

50.60 through 51.78 MHz
50.62 is the 6-meter packet calling frequency

2 meters

Every 20 kHz from 144.91 through 145.09 MHz. 145.01 is the 2-meter packet calling frequency. Packet can also be found between 145.50 and 145.80 MHz. The national Automatic Packet Reporting System (APRS) frequency is 145.79 MHz.

222 MHz and Up

Packet can be found on the 222, 440, 902 and 1240-MHz bands, but bulletin boards and live conversations are sporadic. Most of the activity on the higher bands is in the form of backbone links that pass traffic between bulletin boards and nodes. Avoid these backbones; they're not intended for individual user access.

You may also encounter 9600-baud TCP/IP packet activity on the 440-MHz band in many areas.

ding!) with microwave equipment. But if you decide to tread the airwaves beyond 9600 baud, you'll need special radios and TNCs. You'll definitely be out of the plug-and-play universe!

The TNC

Setting up a TNC isn't complicated. Just read the manual carefully and you'll soon be in business.

Let's talk *parameters*. TNCs can be tailored to match your particular operating conditions by altering the parameters stored in their memories. By changing these parameters you can adjust the time of day setting, change the transmit/receive switching delay, turn your personal mailbox on and off, etc. Your TNC manual will describe all these parameters in detail.

The good news is that you don't have to fiddle with the parameters very much to get on the air. All TNC's come loaded with so-called *default* settings It's best to leave these settings alone until you become more familiar with packet. They're usually adequate to get you up and running.

The most important parameter that you *must* change, however, is the one that contains your call sign. Once you've fired up your TNC and you're sure it's talking to your computer (see the sidebar, "The Gibberish Syndrome"), you must enter your call sign. Just type:

MYCALL <your call sign>

For example, **MYCALL KD4XYZ** tells the TNC to use the call sign KD4XYZ. What could be easier?

I should mention that it's possible for some computers to function as TNCs through the use of specialized software. All that's required is a simple outboard circuit to act as the interface between the computer and your radio. *DIGICOM>64* is a software-based TNC system for Commodore computers. *BayCom* (and similar systems such as

The Gibberish Syndrome

You turn on your computer, fire up your terminal software, switch on your new TNC . . . and your heart sinks like a stone. You expect some sort of greeting from the nifty little box, but instead you see:

alsdkfjalskefalsdfjoijowf

asdlkfjqw0fj0924maksjdf092193874las

@OU#lkjl2kj4d

It's the gibberish syndrome!

Pick your shattered hopes off the floor. Things are not nearly as bad as they seem. The gibberish syndrome is caused by a simple miscommunication between your TNC and your terminal software. Your software expects the TNC to send data at a specific rate, but the TNC is zipping along at an altogether different speed. The solution is to change the terminal software's baud rate.

Every terminal program allows you to change the baud rate. I've yet to find one that doesn't. Read your software manual and it will tell you how. If you don't have a manual handy, look for a function called **SETTINGS** or something similar. In the terminal program that *Windows* provides, for example, you can change the baud rate by accessing the **SETTINGS** area, then **COMMUNICATIONS**.

Your TNC manual will specify the TNC's *factory default* baud rate. That's the data rate that was set when the TNC left the factory. Adjust your software to match this rate and you should be in business.

Some TNCs include an *autobaud* routine that activates when you switch on the TNC for the first time. The TNC sends a short sentence repeatedly at various baud rates. As you're watching your monitor, you see the usual gibberish. Then, suddenly, you see . . .

PRESS (*) TO SET BAUD

. . . or something like that.

The TNC has just sent a message to your computer at a rate that matched your terminal software. If you quickly press *****, the TNC will stop the autobaud routine and store that data rate in its memory. The next time you switch on your TNC, it will communicate at the correct data rate.

By adjusting a TNC parameter (see your TNC manual), you can change the data rate to whatever you desire. The best advice is to communicate at the highest speed your TNC and computer will handle. For most applications, set your software and TNC at 9600 baud. *Don't get confused, though!* This setting has *nothing* to do with the speed at which the TNC sends and receives data through your radio.

BayPac) is designed for IBM PCs and compatibles. These shortcut systems are not as versatile as stand-alone TNCs, but they'll get you on the air quickly and inexpensively (see Table 3-1).

When shopping for a TNC, you must consider whether you want to operate at 1200 baud, 9600 baud, or both. You can purchase TNCs that function at 1200 or 9600 baud *only*. The alternative is to buy a dual-speed TNC that operates at 1200 *and* 9600 baud. The dual-speed boxes are somewhat more expensive, but you get the best of both worlds.

And if you think you'll want to dabble in digital modes other than packet, you can invest in a *multimode controller*. A multimode controller combines a packet TNC with other circuitry and software that allow you to operate the shortwave digital modes such as RTTY, AMTOR, PACTOR and so on. As you might guess, multimode control-

A potpourri of packet TNCs. From top to bottom, left to right: AEA PK-96, MFJ-1270C, Kantronics KPC-9612 and the PacComm TNC/NB96. All four TNCs are capable of 1200 *and* 9600-baud operation.

Multimode controllers such as these give you RTTY, AMTOR, PACTOR, CW and other modes *in addition* to packet.

lers can be a bit pricey. They start at about $300 and go as high as $800, depending on how many bells and whistles you desire.

If your computer has a standard serial port, you can buy a 9 or 25-pin serial cable at Radio Shack and connect your TNC accordingly. That only takes a few seconds. The fun begins when its time to connect the TNC to your radio. If you're handy with a soldering iron and can read electrical diagrams, you can wire your own TNC-to-radio connector. Your TNC manual is a big help in this department.

If the thought of manufacturing a radio cable makes your skin crawl, I suggest that you contact MFJ Enterprises, Box 494, Mississippi State, MS 39762, tel 601-323-5869. They sell prewired radio cables that work with several TNC models and a number of radios. Most of these cables connect the TNC to the microphone jack, so they're only useful for 1200-baud packet. For 9600 baud, you'll have to use the radio's data jack. If the radio doesn't have a data jack, find a friend who can make the internal connections for you.

LET'S CONNECT!

The easiest way to get started in packet is to make a *connection* to another station and have a live (real-time) chat. A TNC is connected to another station when it exchanges the initial protocols and begins the error-checking routine described earlier. If you have a friend who's active on packet, make an appointment to get on the air together and have a chat. Otherwise, monitor a frequency and watch for the call signs of stations you may wish to contact.

Monitoring with your TNC is easy. Consult your TNC manual and I bet you'll find a command called *MONITOR*, *MCOM* or something similar. By invoking this command your TNC will display the contents of every packet it hears. Your TNC also maintains a list of stations it has heard. Just enter the command *MH* and you'll see this list.

To start a packet conversation, you have to issue a *connect request*. In simple terms, your TNC will attempt to contact the other station and "request" a connection. Use **CONTROL-C** (or whatever key combinations your software requires) to place your TNC in the *command* mode. (That's when you see the **cmd:** prompt on your screen.) Everything you enter from the keyboard will now be interpreted as a command to be executed. You don't have to enter the full command. Just an abbreviation will do in most cases. For the sake of clarity, however, I'm going to show the complete command. Type carefully! If the TNC doesn't understand your request, you'll see "EH?", "HUH?" or a similar response!

cmd: CONNECT WB8ISZ <CR>

(Throughout this chapter we'll use <CR> to represent the **ENTER** or **RETURN** key on your keyboard.)

You've just asked your TNC to contact WB8ISZ. Watch your transceiver. It should begin transmitting short packet bursts automatically. (If another station is transmitting, your rig will not be keyed until the other transmission stops.) When the connection is finally established, you'll see the following:

*** CONNECTED to WB8ISZ

Congratulations! You're officially packet-active! You'll notice that the cmd: prompt has disappeared. This means that your TNC has left the command mode and has entered the *converse* mode. Everything you type will now be interpreted as text to send to WB8ISZ rather than commands for your TNC.

Flashing Lights

The glowing array of LEDs that appears on the front panels of most TNCs can intimidate first-time users. This is complicated by the fact that their abbreviated labels are often cryptic at best. Let's take a moment to decipher the puzzle of the flashing lights.

XMIT, SEND or PTT: This indicator glows whenever your TNC sends a packet. It is a red indicator on most TNCs.

RCV or DCD: Whenever your TNC is receiving packet data, this LED will glow. Unless you have your TNC in the *full duplex* mode (rarely used), it will not transmit while this LED is lit.

CON: Once you've established a *connection* to another packet station, this LED

will switch on and remain on.

STA: This LED indicates the status of outgoing packets. It glows when the TNC has a packet ready to send, and winks out when its safe arrival is confirmed. If the **STA** indicator remains on for a long period of time, there may be a problem with the link between you and the other station. Your TNC is sending packets, but their arrival confirmations are not coming back.

MAIL: If your TNC has the ability to store incoming mail, then this LED will glow (or blink) to indicate that you have mail waiting.

PWR or POWER: This LED simply lets you know that your TNC is ready for action.

The rest is up to you. The conversation proceeds like any other, except that you're typing rather than talking or tapping a key. It's important to let the other station know when you're finished with a thought and awaiting his or her response. Without some method to control the "flow" of the conversation, a packet conversation can get pretty confusing! Most packeteers use "K" or ">>" to mean "over." For example:

Hello, Dave. My name is John and I live in Celina >> <CR>

Hi, John! Nice signal here in Dayton. How long have you been on packet? >> <CR>

If you have an outside antenna and a reasonable amount of output power (20 or 30 W), you can enjoy a number of direct, point-to-point contacts. But what if you attempt to make a connection and the other station can't hear you? Your TNC will try to establish communication, but after a certain number of attempts (usually 10) you'll see:

Retry count exceeded

*** DISCONNECTED

This is TNC's way of saying, "I give up!" It looks like you need an intermediate station to act as a relay, doesn't it? That's where *nodes* and *digipeaters* come in!

Packet Nodes and Digipeaters

Many hams who are new to packet confuse nodes and digipeaters with the voice repeaters we discussed in Chapter 2. There are similarities (they both repeat), but there are major differences as well. . .

- Voice repeaters listen on one frequency and transmit on another. Most nodes and digipeaters listen and transmit on the *same* frequency. The exceptions are nodes that act as links to other stations via UHF or VHF *backbone* systems, or nodes that function as *gateways* to HF packet frequencies or the Internet.
- Voice repeaters relay *every* signal heard on the input frequency. Nodes and digipeaters relay only packets addressed specifically to them.
- Voice repeaters are *real-time* devices. They listen and transmit simultaneously. Nodes and digipeaters receive packets, store them momentarily and then retransmit.
- Only one station at a time can use a voice repeater. Several packet stations can use a node simultaneously.

In the early days of amateur packet, nodes didn't exist. If you needed a relay you had to use a digipeater. Every second-generation TNC (called a *TNC2*) had the ability to function as a digipeater, so any station could act as a relay for another. (This is still true today, although it's not a common practice.) Some amateurs established powerful digipeaters by placing TNCs, transceivers and antennas at excellent operating sites. These allowed hams in poor locations to extend their coverage over wide areas.

The big problem with digipeaters was that they weren't "intelligent." They shuffled data from one station to another and nothing else. If you connected to a station through several digipeaters, every bit of information had to travel from one end of the path to the other (see Figure 3-4). The more digipeaters involved in the path, the greater the possibility of failure.

When the first nodes hit the packet scene, the change was startling. All you had to do was get the packet to the node and the node took responsibility for relaying the packet to its destination. For example, if the receiving station received the packet

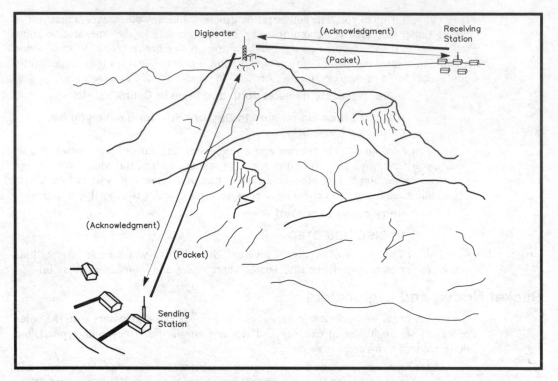

Figure 3-4—Digipeaters are packet relays that you can use to bridge the gaps between you and another station. However, every bit of data must travel back and forth between the sending station, the digipeater and you. That leaves many opportunities for errors.

error-free, its acknowledgment only had to travel to the node—not all the way back to your station (see Figure 3-5)! By eliminating end-to-end acknowledgments, efficiency improved dramatically.

In our previous example, you could have used a node to make the connection to WB8ISZ. Let's call our node WA6ZWJ-5. You begin by establishing a connection to the node.

cmd: CONNECT WA6ZWJ-5 <CR>

If the node acknowledges your request, you'll see:

*** CONNECTED to WA6ZWJ-5

You're connected to the node. Now you can use the node to reach WB8ISZ. Simply enter:

CONNECT WB8ISZ <CR>

You've asked the node to contact WB8ISZ for you. With any luck you'll be rewarded with:

*** CONNECTED to WB8ISZ

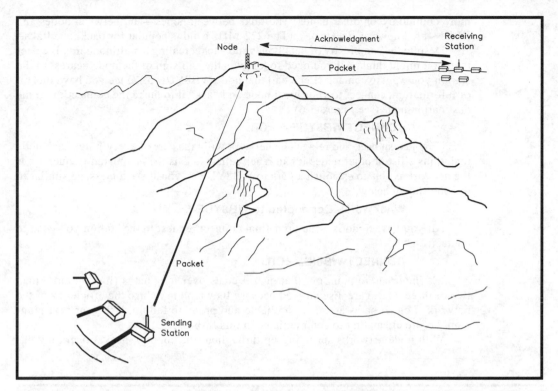

Figure 3-5—Packet *nodes* are a huge improvement over digipeaters. Once the node receives your packet error-free, it "takes responsibility" for relaying it to the receiving station. The acknowledgments for each packet go back to the node *only*, not all the way back to your station. The fewer relays involved, the less chances of errors.

By the way, you're probably wondering why the call sign of the node in our example has a "-5" following the last letter. That's called a *secondary station identifier*, or *SSID*. It's used to separate one function from another at the same station location. WA6ZWJ might also have a bulletin board, for example. If so, he may choose to call it WA6ZWJ-4. By using the full call sign and the proper SSID, you can connect to the station you desire (the node or BBS, in this case).

In addition to their call signs, packet nodes can also be identified by their *aliases*. Don't confuse them with the long lists of names you see on FBI "wanted" posters. Aliases come in handy when you can't seem to remember the call sign of the node you want to reach. They often indicate the location of the node, or the name of the group that operates it. For example, the alias of WA6ZWJ-5 might be RIVER—an abbreviation for Riverside. You can command your TNC to connect to WA6ZWJ-5 or RIVER. The connection will be made regardless of whether you use the call sign or the alias.

If the distance between you and another station is beyond the capability of one node to handle alone, you can link several nodes together to form a multinode path. The nodes pass information by communicating with each other on the same band or, more commonly, on other bands. The links between nodes—and between nodes and BBSs—are known as *backbones*. (The 222-MHz band is popular for backbone links.)

Despite the complexity of backbones and such, creating a multinode link is easier than you might think. All you need to know is the call sign of the node nearest to the station you want to contact. (Let's call that node WB8YUE-5.) Once you have this bit of information, connect to your local node and "ask" it to make the connection to the destination node (see Figure 3-6).

CONNECT WB8YUE-5 <CR>

Now just sit back and relax. The network will take care of everything. Each node maintains a list of other nodes it can reach reliably. This list is updated frequently. If the network is able to establish a path to WB8YUE-5, you'll see a message similar to the one shown below.

WA8ZWJ-5> Connected to WB8YUE-5

All you have to do is issue your final connect request to the station you wish to contact:

CONNECT WB8ISZ <CR>

It's important to point out that packet chats over long paths (linking more than two or three nodes) are discouraged because they tend to reduce the efficiency of the network. Longer paths are also unreliable and prone to failure. Use your best judgment before attempting to communicate in this fashion.

With node networks springing up throughout the country, packet operating has

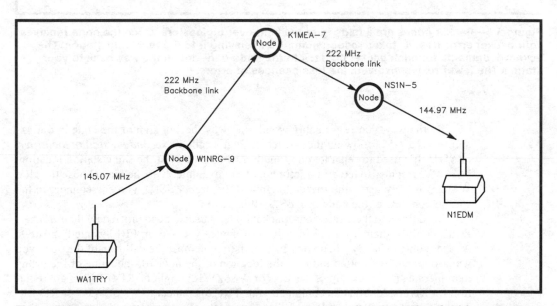

Figure 3-6—You can link several nodes together to reach a distant station. In this example, WA1TRY connects to the W1NRG-9 node on 2 meters. W1NRG-9 relays the data via a 222-MHz backbone link to the K1MEA-7 and NS1N-5 nodes. The destination station, N1EDM, finally receives the data on 2 meters from the NS1N-5 node.

become easier than ever before. Digipeaters still exit, but nodes now dominate most packet networks. Unless you live in a remote area, there is probably a node you can use to connect to many other stations—such as bulletin boards!

PACKET BULLETIN BOARDS

Bulletin boards (BBSs) form the hubs of most VHF packet networks. They're electronic warehouses for the bulletins and mail that flow through the packet system. By connecting to a BBS, you'll be able to read general bulletins or send mail to other packet-active hams.

Most BBSs are operated by clubs or private individuals. It takes a fair amount of time and money to maintain such a system. The system operator—or *SysOp*—is the person who calls the shots on any BBS. After all, it's his time and, in many cases, his money! The worldwide packet network depends upon the dedication of SysOps. When you consider that this is an all-volunteer effort, it is amazing that the system functions as well as it does.

Connecting to a Bulletin Board

Connecting to a BBS is the same as connecting to any other packet station. Monitor the packet frequencies in your area and watch for bulletin board activity. When you find one, make your connect request and have fun! Let's pretend we've discovered the W1NRG-4 BBS.

cmd: CONNECT W1NRG <CR>

We'll also presume that you're close enough to connect directly. If not, you'd use a node to bridge the gap.

*** CONNECTED TO W1NRG

Good! You've made the connection. Now the BBS will send its sign-on information to you. Let's try a couple of the most popular functions. Would you like to see some of the bulletins available on the BBS today? The *L*, or *LIST*, command will do the trick (see Figure 3-7).

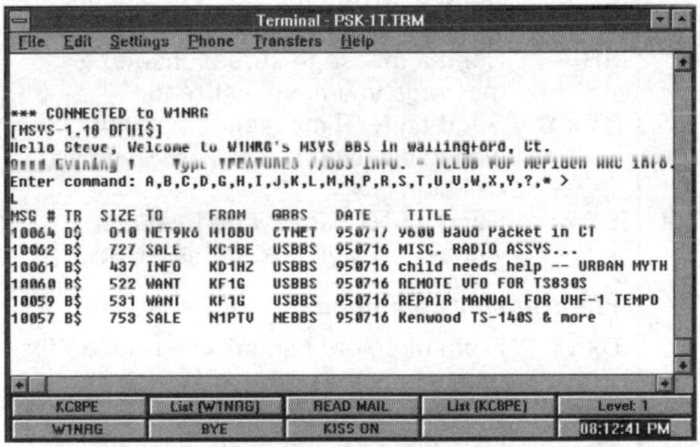

You can read any of these messages by sending the letter *R* followed by the message number. Sending "R 743" allows you to read message number 743 concerning an HF rig for sale.

If you're new to the BBS and you send the *L* command, you're likely to get a huge, long list that will go on and on and on . . . You might want to try some variations on the list command to be a little more selective. For example, *LB* will present you with a list of bulletins *only*. A list of the most common BBS commands—including all the *list* variations—is shown in Table 3-3.

Figure 3-7—After you connect to a packet BBS, you can send the List command (L) to see a list of all new messages. Notice that in this example we're using the terminal program provided in Microsoft *Windows* to communicate with the TNC.

Table 3-3
Common PBBS Commands

General Commands:

B	Log off PBBS.
Jx	Display call signs of stations recently heard or connected on TNC port x.
N x	Enter your name (x) in system (12 characters maximum).
NE	Toggle between short and extended command menu.
NH x	Enter the call sign (x) of the PBBS where you normally send and receive mail.
NQ x	Enter your location (x).
NZ n	Enter your ZIP Code (n).
P x	Display information concerning station whose call sign is x.
S	Display PBBS status.
T	Ring bell at the Sysop's station for one minute.

Information commands:

? *	Display description of all PBBS commands.
?	Display summary of all PBBS commands.
? x	Display summary of command x.
H*	Display description of all PBBS commands.
H	Display summary of all PBBS commands.
H x	Display description of command x.
I	Display information about PBBS.
I x	Display information about sation whose call sign is x.
IL	Display list of local users of the PBBS.
IZn	List users at ZIP Code n.
V	Display PBBS software version.

Message commands:

K n	Kill message numbered n.
KM	Kill all messages addressed to you that you have read.
KT n	Kill NTS traffic numbered n.
L	List all messages entered since you last logged on.
L n	List message numbered n and messages numbered higher than n.
L< x	List messages *from* station whose call sign is x.
L> x	List messages addressed *to* station whose call sign is x.

L@ x	List messages addressed for forwarding to PBBS whose call sign is x.
L n1 n2	List messages numbered n1 through n2.
LA n	List the first n messages stored on PBBS.
LB	List all bulletin messages.
LF	List all messages that have been forwarded.
LL n	List the last n messages stored on PBBS.
LM	List all messsages addressed to you.
LT	List all NTS traffic.
R n	Read message numbered n.
RH n	Read message numbered n with full message header displayed.
RM	Read all messages addressed to you that you have not read.
S x @ y	Send a message to station whose call sign is x at PBBS whose call sign is y.
S x	Send message to station whose call sign is x at this PBBS.
SB x	Send a bulletin message to x at this PBBS.
SB x @ y	Send a bulletin message to x at PBBS whose call sign is y.
SP x @ y	Send a private message to station whose call sign is x at PBBS whose call sign is y.
SP x	Send a private message to station whose call sign is x at this PBBS.
SR	Send a message in response to a message you have just read.
ST x @ y	Send an NTS message to station whose call sign is x at PBBS whose call sign is y.
ST x	Send an NTS message to station whose call sign is x at this PBBS.

File transfer commands:

Dx y	From directory named x, download file named y.
U x	Upload file named x.
W	List what directories are available.
Wx	List what files are available in directory named x.
Wx y	List files in directory named x whose file name matches y.

Sending Packet Mail

Would you like to send a message to a packet-active amateur in another city, state or country? If you know the call sign of his local BBS, it's easy! Let's say that you wanted to send a message to WB8ISZ at the WA8ZWJ BBS. Just enter:

SP WB8ISZ @ WA8ZWJ <CR>

This odd-looking line translates to: Send a personal message (SP) to WB8ISZ at (@) the WA8ZWJ bulletin board. The BBS now asks for the subject of the message:

SUBJECT?:

Enter a brief subject sentence. Keep it *very* short.

WORKED MEXICO ON 6 METERS! <CR>

The BBS will respond with something like this:

ENTER MESSAGE. USE CTRL-Z OR /EX TO END MESSAGE

That's your cue to begin entering the text of your message. If the message is going to travel out of your region, keep it as brief as possible. Longer messages travel much slower. When you've entered your message, enter **CONTROL-Z** or **/EX** on a line by itself. You'll soon see the BBS command line again.

Enter command: A,B,C,D,G,H,I,J,K,L,M,N,P,R,S,T,U,V,W,X,?,* >

How long will it take for the message to reach your friend? It depends on a host of factors, including propagation. The packet network is not a commercial system. You can't expect the same level of reliability. Your message may arrive in a few hours, a few days, a few weeks or not at all! Even so, packet mail is surprisingly reliable and efficient. I can usually send messages from Connecticut to Ohio, for example, in about 12 hours.

Sending a Bulletin

Sending a bulletin is similar to sending private mail. The only difference is that you don't have a particular person or destination in mind. Even so, you want the message to appear on every packet BBS in your state (we'll assume you live in Ohio for the moment)

SB ALL @ OHBBS <CR>

This means: Send a bulletin (SB) to everyone (ALL) at (@) every BBS in Ohio (OHBBS).

Hey, why not blast out your bulletin shotgun style by sending it to USBBS (every BBS in the country) or WW (every BBS in the world)? *Please don't do this unless you have an awfully good reason!* The packet network is clogged with bulletins that have no business going to USBBS or WW. I once saw a worldwide bulletin from a ham in Texas who was selling a 100-foot tower. Imagine a Russian packeteer scratching his head over that one! (I wonder how much UPS would charge to ship a 100-foot tower to Moscow?)

Think long and hard before you use USBBS or WW. Ask yourself, "Does every packeteer in the nation or the world *really* need to read my message?" In many cases the answer will be "no." If you have a question or an item for sale, send the bulletin to all the BBSs in your state or region *first*. If you don't get a response within a few weeks, *then* consider USBBS or beyond.

SUBJECT?

Describe your bulletin in as few words as possible.

Wanted: 2-meter SSB radio <CR>

As before, enter your text and end your message. Connect to the BBS a few days later and you may have some mail waiting!

You can certainly use packet bulletins to announce that you have some goodies for sale. However, FCC rules specify that the items must be directly related to Amateur Radio (a transceiver, an antenna, a computer, a monitor and so on). Don't try to sell your car or your house on the packet network!

DX *PACKETCLUSTERS*

Even if you lack privileges on HF, a DX *PacketCluster* has something to offer!

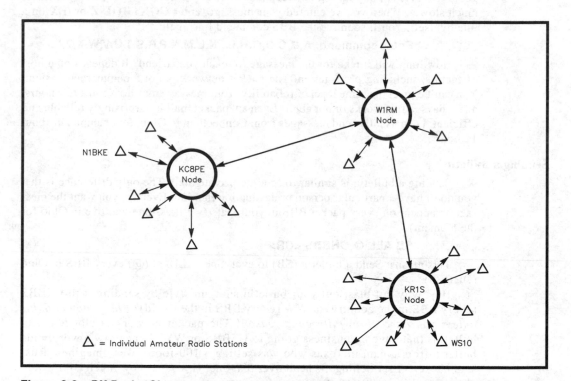

Figure 3-8—*DX PacketCluster* networks are comprised of individual nodes and stations. In this example, N1BKE is connected to the KC8PE node. If he hears a DX station on the air, he'll post a notice—known as a *spot*—which the KC8PE node distributes to all its local stations. In addition, KC8PE passes the information to the W1RM node. W1RM distributes the information and it also passes the data to the KR1S node. Eventually, WS1O—who is connected to the KR1S node—sees the information on his screen.

Not all areas of the country are blessed with *PacketClusters*, but their population is growing rapidly. You may stumble upon one as you're monitoring your local packet frequencies. At first glance it may look like a bulletin board, but it isn't!

A *PacketCluster* is a network of nodes operating under specialized *PacketCluster* software. The network is dedicated to contest and DX activities. Hundreds of stations can connect to a *PacketCluster* network simultaneously (see Figure 3-8). When one station contributes a piece of information, it's seen by everyone else on the network!

You connect to a *PacketCluster* in the same way you'd connect to any other station. However, the information you'll receive will be very different. You can get all sorts of useful information from a *PacketCluster*. How about a list of the latest DX sightings (called *spots*)? Enter: **SHOW/DX <CR>**

7015.5	UL7MG	5-May-1995 0119Z	<W3XU>
14029.3	UD6DFF	5-May-1995 0106Z	<K2LE>
14211.9	J68AJ	5-May-1995 0056Z	<NE3F>
14002.6	4S7WP	5-May-1995 0054Z	<K2LE>
14007.0	ZA1ED	5-May-1995 0048Z	<K2LE>

Now you have a list of the five most recent DX spots along with their frequencies and the times (in UTC) when they were heard. If you're tuning through the bands you may discover another DX station worthy of a spot on the cluster. Go ahead and make a contribution by posting it on the network. You've just heard SV3AQR on 15 meters. The simplest command format would be:

DX 21.250 SV3AQR <CR>

If you don't have HF privileges, keep an eye on your local *PacketCluster* anyway. When significant VHF band openings occur, you'll often hear about it first on the *PacketCluster*.

PacketClusters have many other useful features. You can use the DIR command to see a list of bulletins just as you would on a packet BBS. Using the R (READ) command will allow you to read any bulletin you wish. Unlike packet bulletin boards, however, *PacketClusters* can relay bulletins and messages only within the networks they serve. You can also chat with other hams on the network through the use of the TALK command.

TALK WS1O Hello Brian. What's up?

Within seconds, depending on the size of the network and how busy it is, Brian will see my greeting on his screen. This is a handy function, especially if you want to congratulate someone after working a new country, for example. See Table 3-4 for a list of the most common *PacketCluster* commands.

Other Packet Goodies

There is much more to packet radio than BBSs and DX *PacketClusters*. As you explore this corner of the hobby you'll encounter many other wonders guaranteed to astound the imagination.

TCP/IP

If you want to expand your packet horizons, try *Transmission Control Protocol/ Internet Protocol*—otherwise known as *TCP/IP*. Assuming that you already own a

Table 3-4
Common DX *PacketCluster* Commands

ANNOUNCE	Make an announcement.	FI x	Ask the node to find file named x.
A x	Send message x to all stations connected to the local node.	HELP or ?	Display a summary of all commands.
A/F x	Send message x to all stations connected to the cluster.	HELP x	Display help for command x.
		READ	Read message.
A/x y	Send message y to stations connected to node x.	R	Read oldest message not read by you.
A/x y	Send message y to stations on distribution list x.	Rn	Read message numbered n.
BYE	Disconnect from cluster.	R/x y	Read file named y stored in file area named x.
B	Disconnect from cluster.	REPLY	Reply to the last message read by you.
CONFERENCE	Enter the conference mode on the local node.	REP	Reply to the last message read by you.
CONFER	Enter the conference mode on the local node. Send <CTRL-Z> or /EXIT to terminate.	REP/D	Reply to an delete the last message read by you
CONFER/ F	Enter the conference mode on the cluster. Send <CTRL-Z> or /EXIT to terminate.	SEND	Send a message.
		S/P	Send a private message.
		S/NOP	Send a public message.
DELETE	Delete a message.	SET	Set user-specific parameters.
DE	Delete last message you read.	SE/A	Indicate that your computer/terminal is ANSI-compatible.
DE n	Delete message numbered n.		
DIRECTORY	List active messages on local node.	SE/A/ALT	Indicate that your computer/terminal is reverse video ANSI-compatible.
DIR/ALL	List all active messages on local node.		
DIR/BULLETIN	List active messages addressed to "all."	SE/H	Indicate that you are in your radio shack.
DIR/n	List the n most recent active messages.	SE/L a b c d e f	Set your station's latitude as: a degree b minutes c north or south and longitude d degrees e minutes f east or west, e.g., SE/L 41 33 N 73 0 W.
DIR/NEW	List active messages added since you last invoked the DIR command.		
DIR/OWN	List active messages addressed from or to you.	SE/N x	Set your name as x.
DX	Announce DX station.	SE/NEED x	Store in database that you need country(s) whose prefix(s) is x on CW and SSB. e.g., SE/NEED XX9.
DX x y z	Announce DX station whose call sign is x on frequency y followed by comment z, e.g., DX SP1N 14.205 up 2.		
		SE/NEED/BAND	Store in database that on frequency band(s) =(x)y x, you need country(s) whose prefix(s) is y, e.g., SE/NEED/ AND=(10)YA.
DX/a x y z	Announce DX station whose call sign is x on frequency y followed by comment z with credit given to station whose call sign is a, e.g., DX/K1CC SP1N 14.205 up 2.	SE/NEED/x y	Store in database that in mode x (where x equals CW, SSB or RTTY), you need country(s) whose prefix(s) is y, e.g., SENEED/RTTY YA.
FINDFILE	Find file.		

Command	Description
SE/NEED/ x/BAND	Store in database that in mode x (where =(y)z x equals CW, SSB or RTTY) on frequency band(s) y, you need country(s) whose prefix(s) is z, e.g., SE/NEED/ RTTY/BAND =(10) ZS9.
SE/NOA	Indicate that your computer/ terminal is not ANSI-compatible.
SE/NOH	Indicate that you are not in your shack.
SE/Q x	Set your QTH as location x.
SHOW	Display requested information.
SH/A	Display names of files in archive file area.
SH/B	Display names of files in bulletin file area.
SH/C	Display physical configuration of cluster.
SH/C x	Display station connected to node whose call sign is x.
SH/CL	Display names of nodes in clusters, number of local users, number of total users and highest number of connected stations.
SH/COM	Display available Show commands.
SH/DX	Display the last five DX announcements.
SH/DX x	Display the last five DX announcements for frequency band x.
SH/DX/n	Display the last n DX announcements.
SH/DX/n x	Display the last n DX announcements for frequency band x.
SH/FI	Display names of files in general files area.
SH/FO	Display mail-forwarding database.
SH/H x	Display heading and distance to country whose prefix is x.
SH/I	Display status of inactivity function and inactivity timer value.
SH/LOC	Display your station's longitude and latitude.
SH/LOC x	Display the longitude and latitude of station whose call sign is x.
SH/LOG	Display last five entries in cluster's log.
SH/LOG n	Display last n entries in cluster's log.
SH/M x	Display MUF for country whose prefix is x.
SH/NE x	Display needed countries for station whose call sign is x.
SH/NE x	Display stations needing country whose prefix is x.
SH/NE/x	Display needed countries for mode x where x equals CW, SSB, or RTTY.
SH/NO	Display system notice.
SH/P x	Display prefix(s) starting with letter(s) x.
SH/QSL x	Display QSL information for station whose call sign is x.
SH/S x	Display sunrise and sunset times for country whose prefix is x.
SH/U	Display call signs of stations connected to the cluster.
SH/V	Display version of the cluster software.
SH/W	Display last five WWV propagation announcements.
TALK	Talk to another station.
T x	Talk to station whose call sign is x. Send <CTRL-Z> to terminate talk mode.
T x y	Send one-line message y to station whose call sign is x.
TYPE	Display a file.
TY/x y	Display file named y stored in file area named x.
TY/x/n y	Display n lines of file named y stored in file area named x.
UPDATE	Update a custom database.
UPDATE/x	Update the database named x.
UPDATE/x/ APPEND	Add text to your entry in the database named x.
UPLOAD	Upload a file.
UP x	Upload a file named x.
UP/B x	Upload a bulletin named x.
UP/F x	Upload a file named x.
WWV	Announce and log WWV propagation information.
W SF=xxx, A=yy, K=zz,a	Announce and log WWV propagation information where xxx is the solar flux, yy is the A-index, zz is the K-index and a is the forecast.

packet station, the additional cost to run TCP/IP is . . . *nothing*. Nada. Zip. Nil.

If you've been fortunate enough to surf the Internet, you've already dabbled in TCP/IP. (Although if you're using a *shell* program that makes the Internet more user friendly, many of the nuts and bolts of TCP/IP are transparent to you.) As complicated as it may sound, TCP/IP is simply a group of packet *protocols* that make it possible for the Internet to shuffle information throughout the world.

TCP/IP was pretty much confined to the Internet and Unix-type computer systems until Phil Karn, KA9Q, decided to adapt it for use on ham networks and IBM PCs. The operating system he created was originally called *NET*. In time he rewrote the software and it debuted to the ham populace as *NOSNET*, or just *NOS* for short (*Net Operating System*). Not long thereafter, *NOS* was adapted to run on other computers.

Of course, hams can't resist the urge to tinker. Talented programmers followed Phil's lead and created their own versions of *NOS*. That's why you'll find so many available. There's *GRINOS*, *JNOS*, *TNOS*, *MFNOS* and so on. All of these *NOS* versions use the same TCP/IP protocols. Even the original Unix jargon remains.

But why go to so much trouble? After you get the software running and the hardware perking along, what do you have?

Simply this: an Amateur Radio version of the Internet.

The Attraction of TCP/IP

When a group of packet stations decide to adopt TCP/IP as a way of communicating among themselves, the resulting network looks a lot like the Internet. We call it the *AMPRNET—Amateur Packet Radio Network*. It operates in much the same fashion as the Internet and, as I've already mentioned, it uses the same protocols. The primary difference is that amateur networks are usually much slower and more limited in scope.

Some TCP/IP networks are very small—just a loose group of dedicated enthusiasts who enjoy swapping information. At the opposite end of the scale, there are vast networks that offer coverage to packeteers in several states. They achieve this coverage through the use of special TCP/IP nodes (often called *switches*) that act as relays.

Several of these large networks include *gateways* to transfer information to and from the Internet. The same gateways can also function as *wormholes*—Internet pipelines that link distant sections of large networks.

But what makes TCP/IP so special? Here are just a few of the high points . . .

Mail—TCP/IP mail is sent from station to station through the network. There are no BBSs involved. You need only prepare a message and leave it in your own TCP/IP "mailbox." Within seconds your computer will attempt to make a connection to the target station and deliver the message. If the target station isn't on the air, your computer holds the message and keeps trying. When you check your computer and see that the message is no longer in the mailbox, you can be sure it has arrived safely at the other station.

File Transfers—You can use the file-transfer protocol (FTP) to easily transfer files (games, images, or whatever) to any station on the network.

Data Handling—Rather than spewing data at random intervals like standard packet, TCP/IP stations take the smart approach and automatically *adapt* to network delays. As the network slows down, TCP/IP stations sense the change and lengthen their transmis-

Your Own IP Address

Before you can get on the air with TCP/IP, you need your own *AMPRNET* IP address. These addresses are assigned by the volunteer coordinators listed below. This list is subject to change without notice. If you have Internet ftp capability, you'll find updated *AMPRNET* coordinator lists at **ftp.ucsd.edu**.

AK	John Stannard, KL7JL
AL	Richard Elling, KB4HB
AR	Richard Duncan, WD5B
AZ	David Dodell, WB7TPY
CA, Antelope Valley/Kern County	Dana Myers, KK6JQ
CA, Los Angeles —San Fernando Valley	Jeff Angus, WA6FWI
CA, Orange County	Terry Neal, AA6TN
CA, Sacramento	Bob Meyer, K6RTV
CA, Santa Barbara/Ventura	Don Jacob, WB5EKU
CA, San Bernardino and Riverside	Geoffrey Joy, KE6QH
CA, San Diego	Brian Kantor, WB6CYT
CA, Silicon Valley —San Francisco	Douglas Thom, N6OYU
CO (north)	Fred Schneider, K0YUM
CO (south)	Bdale Garbee, N3EUA
CO (west)	Bob Ludtke, K9MWM
CT	Bill Lyman, N1NWP
DC	Richard Cramer, N4YDP
DE	Butch Rollins, NF3F
FL	Brian Lantz, KO4KS
GA	Doug Reed, N3AIA
HI and Pacific islands	John Shalamskas, KJ9U
IA	Ron Breitwisch, KC0OX
ID, eastern WA	Steven King, KD7RO
IN	Jacques Kubley, KA9FJS
IL (central and southern)	Chuck Henderson, WB9UUS
IL (northern)	Ken Stritzel, WA9AEK
KS	Dale Puckett, K0HYD
KY	Allan Dayton, N0KFO
LA	James Dugal, N5KNX
MA (center and eastern)	Gordon LaPoint, N1MGO
MA (western)	Bob Wilson, KA1XN

MD	Howard Leadmon, WB3FFV
ME	Carl Ingerson, N1DXM
MI (upper peninsula)	Thomas Landmann, N9UDL
MI (lower peninsula)	Jeff King, WB8WKA
MN	Andy Warner, N0REN
MO	Stan Wilson, AK0B
MS	John Martin, KB5GGO
MT	Steven Elwood, N7GXP
NC (eastern)	Mark Bitterlich, WA3JPY
NC (western)	Charles Layno, WB4WOR
ND	Steven Elwood, N7GXP
NE	Mike Nickolaus, NF0N
NH	Gary Grebus, K8LT
NJ (northern)	Dave Trulli, NN2Z
NJ (southern)	Bob Applegate, WA2ZZX
NM	J Gary Bender, WS5N
NY (eastern)	Bob Bellini, N2IGU
NY (western)	Dave Brown, N2RJT
New York City and Long Island	Bob Foxworth, K2EUH
NV (southern)	Earl Petersen, KF7TI
NV (northern)	Bill Healy, N8KHN
OH	John Ackerman, AG9V
OK	Joe Buswell, K5JB
OR	Ron Henderson, WA7TAS
OR (northwest and Vancouver, WA)	Tom Kloos, WS7S
PA (eastern)	Doug Crompton, WA3DSP
PA (western)	Bob Hoffman, N3CVL
PR	Karl Wagner, KP4QG
RI	Charles Greene, W1CG
SC	Mike Abbott, N4QXV
SD	Steven Elwood, N7GXP
TN	Jeff Austin, K9JA
TX (northern)	Jack Snodgrass, KF5MG
TX (southern)	Kurt Freiberger, WB5BBW
TX (western)	Rod Huckabay, KA5EJX
UT	Matt Simmons, KG7MH
VA	Jim DeArras, WA4ONG
VA (Charlottesville)	Jon Gefaell, KD4CQY
VI	Bernie McDonnell, NP2W
VT	Ralph Stetson, KD1R
WA (eastern)	Steven King, KD7RO
WA (western)	Bob Donnell, KD7NM
WI	Thomas Landmann, N9UDL
WV	Rich Clemens, KB8AOB
WY	Reid Fletcher, WB7CJO

sion delays accordingly. (They don't transmit as often.) As the network speeds up, the TCP/IP stations shorten their delays to match the pace (they transmit more frequently). This kind of intelligent network sharing virtually guarantees that all packets will reach their destinations with the greatest efficiency the network can provide.

Direct Addressing—Every TCP/IP station in your network has an address. These *IP* addresses are assigned by volunteer coordinators throughout the country (see the sidebar, "Your Own IP Address").

My address, for example, is 44.88.4.35. Reading from left to right, 44 tells you that this is the address of an Amateur Radio TCP/IP station (rather than a nonham Internet site). The 88 designates my part of the New England regional network (Connecticut). The 4 gets even more specific, pointing to my little corner of the state (known as a *subnet*). Finally, the 35 is my unique address in the subnet.

But you don't need to know the addresses of your fellow TCP/IP users! All of this information is contained within your *NOS* software in a file called *DOMAIN.TXT*. When I want to send a message to, say, WS1O, I simply address the message to WS1O. When it's time to send the message, *NOS* will search through DOMAIN.TXT and locate WS1O. When it does, it'll use the IP address it finds there.

By analyzing the IP address, the switches and other stations probe through the network to create a link to WS1O. Once the "circuit" is established, the data flows and the message is delivered. All of this takes place without you lifting a finger!

Many versions of *NOS* contain DOMAIN.TXT. It may be a small file that includes only the users in a particular area. On the other hand, it could be the granddaddy of all DOMAIN.TXTs that lists every amateur TCP/IP user in the world! If your *NOS* doesn't have a DOMAIN file, you'll need to find a copy. See the sidebar, "Getting Started in Four Steps."

Multitasking—With TCP/IP you can do several things simultaneously. For example, you can send a message, receive a message and transfer a file—all at the same time!

Flexibility—*NOS* provides an excellent platform for developing new protocols that will dramatically expand the capabilities of the AMPRNET. Some versions already support gopher and an amateur version of the World Wide Web is on the way.

Ping! Is Anyone Home?

Is your buddy on the air this afternoon? Should you bother trying to get a message to him, or transfer a file to his computer? TCP/IP gives you an easy way to find out. Using WS1O as our example, I can determine if his station is active by issuing a *ping*.

ping ws1o

My station sends a response request to WS1O. It weaves through the network until it arrives at his station. His computer responds and I receive a reply, along with the time (in milliseconds) that it took for my ping to get there and back.

44.88.4.23: echo reply 7000 ms

All this jargon basically tells me that WS1O is on the air (his address is shown) and that my ping required 7000 milliseconds (7 seconds) to make the trip.

Give Him a Finger

Let's say that you want to find out more about WB8IMY without bothering to send a message and wait for my reply. The quick way to do this is through the *finger* command. It's a terrific feature for nosy packeteers!

finger wb8imy@wb8imy

Assuming that you have access to my TCP/IP network, your finger request will travel to my station. Like many TCP/IP operators, I have a special text file that provides a brief rundown on my station. When my computer receives a finger request, this file is sent automatically.

Connect with Telnet

The *telnet* function allows you to do keyboard-to-keyboard work. You can connect to another station on the network and access his or her mailbox, get a list of

---

Getting Started on TCP/IP in Four Steps

1. If you don't own a packet station now, you'll need to build one. All that's required is a computer, a 2-meter FM transceiver and a *terminal node controller*, or *TNC*. The TNC must be capable of operating in the KISS mode (*Keep It Simple, Stupid*—no kidding!). All TNCs made within the last 8 years or more include KISS.

2. Get a version of *NOS*. If you can determine the version that's most popular on your network, it's a good idea to use the same. It's much easier for the locals to help you if you're using similar software (see Step 4).

You'll find *NOS* software on many on-line services such as CompuServe (in the *HamNET* forum) and on a number of ham-oriented computer bulletin boards. (Yes, the ARRL BBS has several *NOS* versions available. Call 060-594-0306 and download one of them.) If you have Internet capability, you can ftp many versions of *NOS* from **ftp.ucsd.edu**. *NOS* is also offered for a nominal price by TAPR (Tucson Amateur Packet Radio). For a complete list, send a self-addressed stamped envelope to: TAPR, 8987-309 Tanque Verde Rd, #337, Tucson, AZ 85749-9399.

3. Get an IP address. Contact one of the coordinators shown in the sidebar "Your Own IP Address." Some coordinators may prefer to issue a temporary "test" address at first. If you find that you like TCP/IP, they may ask that you send a message to them via the network and request a permanent address. They're not making you jump through hoops for their enjoyment. It's just that they don't want to assign permanent addresses to hams who are not going to be active on the network for the long term.

4. Install your *NOS* and get on the air. I *strongly* recommend that you get outside help from an experienced TCP/IP operator for this step. You must configure your AUTOEXEC.NOS file with your IP address, routing parameters (who will relay your packets?) and several other bits of information. This can be a frustrating, time-consuming exercise if you don't know what you're doing!

Local TCP/IP operators are your best resource, by far. They'll be able to help you get your station on the air in the shortest time. They'll also have the most up-to-date version of DOMAIN.TXT for your particular network. TCP/IP packeteers are almost evangelical in their quest for new converts, so you'll likely find several willing to assist. If you're already active on standard packet, drop a message on your local bulletin board and ask for assistance. TCP/IPers usually check the local BBSs regularly.

If you must install *NOS* by yourself, pick up a copy of *NOSintro* by Ian Wade, G3NRW. This book will supply much of the information you need. See your favorite dealer or contact the ARRL directly.

---

stations heard and so on. If the other operator is present at the keyboard, you can even enjoy a "live" conversation.

Many TCP/IP Internet gateways allow you to use telnet to connect to an Internet Relay Chat (*IRC*). This is the equivalent of a conversational food fight in cyberspace! (Everyone is talking to each other at once.) If you don't mind the confusion, you'll enjoy tapping into an IRC.

TCP/IP Q & A

Q: Can I use *any* TNC with TCP/IP?

A: As long as it includes the *KISS* mode, yes. (Most TNCs do.) To put your TNC in KISS, you may have to enter the command **KISS ON** followed by **RESET**. To get out of KISS, some TNCs require you to hold the **ALT** key and type 192. Hold **ALT** again and type 255. Release the **ALT** key and then press it one more time while typing 192. These procedures vary, so consult your TNC manual.

Q: Does it have to be a 9600-baud TNC?

A: No. You'll find plenty of 1200-baud TCP/IP activity, primarily on 2 meters. If you can upgrade to a 9600 baud station, so much the better. TCP/IP really shines at high data rates! Most 9600-baud TCP/IP activity seems to take place on the 440-MHz band.

Q: What kind of computer do I need?

A: I recommend an AT-class IBM PC or compatible such as a 286, 386 or 486. Some versions of *NOS* will run on old XTs, but others won't. Some TCP/IP operators also enjoy using Apple Macintosh machines.

Q: Do I have to leave my computer on 24 hours a day to receive mail?

A: If you can leave your TCP/IP station running continuously, you'll be able to send and receive mail at any time. This is convenient for everyone concerned. Some packeteers buy used PCs (such as relatively cheap 286 machines) and dedicate them to their TCP/IP stations. They just park them in a corner with the other equipment and leave them on day and night.

This approach isn't practical for everyone, though. That's where *POP* comes into play. POP stands for *Post Office Protocol*. If another station on your network is active 24 hours a day, he or she may be willing to act as a depository for your incoming mail—otherwise known as a POP *server*. By activating the POP function in your *NOS* software, your station will automatically contact your POP server and grab any waiting mail. This usually takes place immediately after your station comes on the air.

Q: Can I still connect to my friends on standard (AX.25) packet if I'm running *NOS*?

A: Absolutely! You can still connect to your AX.25 BBS and your friends can connect to you. For example, the version of *NOS* I use allows me to make an AX.25 connect to my local BBS by typing the following command:

c 2m w1nrg

The letter "c" stands for "connect" while "2m" is the designator my *NOS* uses to access my TNC. (In some versions of *NOS*, this might be "ax0.") Finally, "w1nrg" is the call sign of my local packet BBS. When I enter this command, *NOS* will use my TNC to make a standard packet connection to W1NRG.

Some AX.25 packet BBSs even have "ports" to the TCP/IP network. This means that you can use your TCP/IP telnet command to access the BBS without resorting to the command line shown above, and without leaving your TCP/IP network frequency. It's as though the BBS has a split personality—it's a TCP/IP system *and* a standard packet BBS!

The Internet Connection

I'm sitting at the keyboard of a friend's computer. My friend isn't a ham. He's a denizen of the *Internet*, a vast global network that shuffles data between millions of people each day. His link to *cyberspace* is through a private Internet account that he uses as part of his business.

I've been trying to convert my friend to Amateur Radio, but it's been tough sledding. After all, how do you compete with a system that offers reliable, worldwide communication any time of the day or night? As I was pondering this fact, I noticed the gray cord that snaked from his computer to a nearby telephone jack. Maybe *that* was the answer!

I made another Amateur Radio sales pitch while he listened with polite bemusement. "Go ahead, convince me," he said as he gestured at the keyboard.

We accessed his Internet account and I entered:

telnet 44.88.4.35

I just told his computer to establish a link to address 44.88.4.35. The request traveled through the labyrinth of the Internet as it sought its destination. One thing was known for certain: the number "44" in the beginning of the address identified it as belonging to an Amateur Radio TCP/IP packet station.

A thousand miles away, my packet station was faithfully monitoring 145.53 MHz. That's the frequency of the W1EDH TCP/IP packet node (repeater) near my home town. The station belonging to Bill Lyman, N1NWP, was on the same frequency. N1NWP was also linked directly (by wire) to the Internet.

Sure enough, the link request finally arrived at N1NWP via the Internet. His computer "looked" at my IP address and knew exactly what to do. A connect request blasted out on 145.53 MHz and was relayed to my station.

"Wait a minute," my astonished friend said. "We're linked to your house, but you don't have an Internet connection. How is this possible?"

"N1NWP is a packet/Internet gateway station. He's able to connect to my computer by radio over a system that we call *packet radio*. Packet isn't nearly as fast as the Internet, but it has one major advantage: it's *portable*. By using radio links, I can set up an amateur packet station almost anywhere." I pointed to the modem cable. "Try cutting that cable and see how well you can communicate!"

A thin smile appeared. He nodded slowly. "Tell me more about packet radio."

(Once you've got 'em hooked, it's only a matter of reeling them in!)

The Elusive Goal

As amateur packet radio has spread throughout the world, one goal has remained elusive: finding a fast, reliable way to transfer information between distant points. At one time we thought that we could build high-speed RF packet networks encircling the Earth, but this is not likely to happen anytime soon. The cost is outrageous and the required maintenance would be too much for most individuals or groups to bear.

We've been using HF packet to help bridge the gaps, but its performance leaves much to be desired. Other HF digital modes have improved the situation, but they're all at the mercy of the changeable nature of HF propagation. None of them offer the speed that even a 1200 baud VHF/UHF packet link can achieve.

Digital Amateur Radio satellites have been pressed into service to transfer packet mail. The satellites are reasonably dependable, but they are not *geostationary* (stationed at fixed points in the sky from our perspective on the ground). Instead, they orbit the Earth at relatively low altitudes. This means that packet stations have only a few opportunities each day to get data to and from the satellites.

It wasn't long before many hams began to consider *nonamateur* means to reach the same end. That's when they began taking a serious look at the Internet.

Opening the Gateways

An Internet connection is as close as a telephone line. With the proper hardware and software, it isn't difficult to interface amateur packet radio stations to the Internet. These are the gateway stations I mentioned earlier.

Some gateway stations are set up at colleges, universities or businesses. Because the Internet connections already exist, the people in charge allow the gateway operators to use them. Other gateway stations are operated by hams who have Internet access at their homes for personal or business use.

Once these gateway stations transfer packet radio data to the Internet, the information moves at high speeds to almost anywhere in the world. Data also pours out of the Internet through the gateways and ends up on amateur packet networks.

Although it isn't a common-carrier service like the telephone companies, the reliability and speed of the Internet are impressive. You can think of the Internet as a multilane superhighway. The amateur packet network is likened to the slower surface streets. And what about the gateway stations? They're the entrance and exit ramps.

Internet gateway stations have been popping up all over the place in recent years. Hams are using them to act as links between packet networks. For example, a packet bulletin board in New England may pass messages to the West Coast via an Internet gateway, rather than relying on the HF modes or satellites. This is often known as a *wormhole* link. Messages that took days to reach their final destinations arrive within hours or even minutes. Other hams use the gateways to explore on their own, tapping the resources of the Internet from their keyboards.

Using the Gateways

The types of activity you can enjoy through a gateway depend on the gateway operator. Some gateways allow access only by specific amateur stations (such as BBSs) or Internet sites. Others limit individuals to sending and receiving mail, or joining real-time "conferences." Here's a quick rundown of what you might encounter:

- *Electronic Mail*: Some gateways allow you to pass packet mail to and from the Internet. For example, it might be possible to send mail to a nonham who has Internet access at work. Ask the gateway operator to tell you how to create e-mail addresses for mail going to or from the Internet.

  Gateway operators sometimes shy away from offering e-mail capability because of possible legal complications (violations of third-party traffic rules and so on). Others solve the problem by reviewing all e-mail messages that arrive from the Internet before passing them onto the packet network.

- *FTP*: Only a few gateways offer the ability to transfer (ftp) files to and from the Internet. The main problem is the slow data rates on the radio side of the link. Even at 9600 baud, transferring a large file could take a long time and seriously bog down the gateway.

- **Telnet**: This is a fun option if your gateway allows it. Telnet amounts to remote access of a computer. It's possible to use telnet to access an amateur packet station remotely from a computer connected to the Internet. (The same process can work in reverse, too.)

For example, you may be able to access my home station—when it's on the air—from the keyboard of any computer that's connected to the Internet. This includes any Internet telnet access offered by the on-line providers such as America On-Line and others.

- **QSO Bridge**: This is a live, keyboard-to-keyboard roundtable that's on 24 hours a day. If your gateway offers access to the bridge, you can enjoy conversations with hams throughout the world. Most of them are sitting at the keyboards of packet stations just like yours.

AUTOMATIC PACKET REPORTING SYSTEM (APRS)

Imagine staring at a map on your computer monitor. It's a map of your state. Blue lines indicate rivers and other bodies of water. Green lines trace major highways. You press a key and zoom into a particular area of your state. Up and down the river you see clusters of symbols with Amateur Radio call signs. As you watch, your packet TNC receives a transmission . . . and one of the symbols *moves!*

Is this still packet radio? To quote Dorothy from *The Wizard of Oz*, "Toto, I don't think we're in Kansas anymore!"

But if Dorothy had been carrying an Automatic Packet Reporting System in her basket, we'd at least be able to see where she was! We would even be able to follow the path she takes on her way to the Emerald City. All we'd have to do is follow the Yellow Brick . . . er, the computer-generated map (see Figure 3-9).

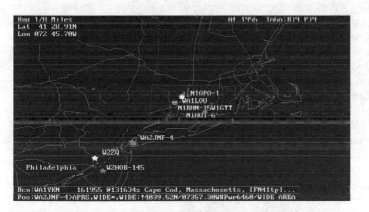

Figure 3-9—Automatic Reporting System (APRS) software displays a map on your computer screen and shows the locations of all other APRS stations on the frequency. If any of these stations change their positions, you'll see the results on the screen right away. You can zoom in for a closer look, or zoom out to see the big picture. In the case of this APRS map, we're looking at Connecticut along with parts of New York and New Jersey.

The Automatic Packet Reporting System (APRS) is the brainchild of Bob Bruninga, WB4APR. APRS exploits the ability of a TNC to transmit *beacon* packets that carry short strings of alphabetical characters or numbers. A beacon is an *unconnected* packet. You can think of unconnected packets as "broadcasts." The information is sent to no one in particular and can be received by anyone. An unconnected packet can be relayed through a node or digipeater if you "tell" your TNC to do so.

The ability to send beacons

has been a feature of TNCs since the earliest days of packet, but two developments in the '90s finally made APRS feasible. The first was the introducing of small TNCs that could operate on battery power. These little boxes could go anywhere without relying on ac power. The second was the debut of compact, affordable *global positioning system* (GPS) receivers. GPS receivers rely on signals from military satellites to determine your position to an accuracy of ± 100 feet. Most of these receivers have ports that allow their data output to be transferred to other devices . . . such as TNCs.

It didn't take long for Bob to realize that he was on to something big. If you could take the position information from a GPS receiver and incorporate it into beacon packets transmitted by a TNC, you could tell everyone on the network *exactly where that GPS receiver was located.* And if the TNC could pass that information along to a computer that could display the position by using a symbol on a map . . . bingo! The Automatic Packet Reporting System was born.

Although APRS' mapping capability was developed to display the movement of hand-held GPS receivers, most features evolved from earlier efforts to support real-time packet communications at special events. Any person in the network, upon determining where an object is located, can move his cursor and mark the object on his map screen. This action is then transmitted to all screens in the network, so everyone gains, at a glance, the combined knowledge of all network participants!

Let's say you're monitoring the movements of rafts during a river race. Each rafter carries a 2-meter FM transceiver, a TNC and a GPS receiver. If your station picks up a transmission from any raft along the river, it will automatically relay the information to everyone else. So, everyone's maps are continually updated with the latest positions of the rafts.

The availability of GPS receiver *cards* is icing on the cake. With a GPS card, TNC and hand-held transceiver stuffed in a cigar box, almost any object can be tracked by packet stations running APRS software. For example, such cigar-box technology was installed in a football helmet for the Army-Navy football game run. Hams with TNCs and APRS software followed the runner as he weaved his way along the course. You can also place these boxes on bicycles for a marathon event, and, of course, in automobiles.

You don't need to buy a GPS receiver to enjoy APRS. All you need is the APRS software and your normal packet TNC. Just determine your latitude and longitude as best you can. Look it up in an atlas, or borrow a friend's GPS receiver just long enough to determine the position of your station. After you feed your position information to the software, your TNC will regularly announce your position to anyone else who is monitoring. You can even use APRS to exchange bulletins and enjoy live conversations with others on the network.

Most APRS activity is on 2 meters, with 145.79 MHz being the popular frequency. If you do purchase a GPS receiver, you'll need a TNC with APRS *firmware*. The good news is that most popular TNCs include the option to add APRS firmware.

The APRS software is distributed as shareware and can be found on many Amateur Radio telephone BBSs, such as the ARRL BBS at 860-594-0306. The software includes maps for most areas of the US. You can also edit and add more detail to the maps. If you don't have access to a telephone BBS, registered copies for IBM PCs are available for $33 from Bob Bruninga, WB4APR, 115 Old Farm Ct, Glen Burnie, MD 21060. A Macintosh version of APRS, MacAPRS, was developed by Keith Sproul, WU2Z, and Mark Sproul, KB2ICI.

Barking at the Moon 4

All generalizations are dangerous . . . including this one.

—*Anonymous*

Look beyond FM voice and packet. What do you see? Is the rest of the VHF and UHF amateur spectrum a trackless desert . . . or a fertile land rich in exciting opportunities?

The answer depends on your expectations. If you expect a nonstop gabfest whenever you switch on the radio, you'll be gravely disappointed. If you expect consistently strong signals and clear communications, you'd better turn to another chapter. But . . .

If you're the type of person who enjoys taking the road less traveled, the type of person who revels in the unknown, this may be one of the most important chapters in the book!

"WEAK SIGNALS"?

If you listen to veteran hams talking about single sideband (SSB) and CW activity on the VHF/UHF bands, you might hear it referred to as "weak signal" operating. Doesn't sound very encouraging, does it? I mean, who wants to have a weak signal? Where's the excitement in weak signals?

The term "weak signal" is a bit misleading. It means that you're often dealing with signals that have traveled great distances, losing much of their energy along the way. Compared to the sledgehammer signals you get from your local repeater, these are weak indeed. In most cases you need beam antennas to concentrate the feeble energy and perhaps a preamplifier to boost the sensitivity of your receiver. SSB and CW are used because these modes are most efficient when you're communicating directly on VHF and UHF. SSB and CW signals are detectable at levels where FM signals can't even be heard.

OUT OF THE WOODWORK

One of the secrets to enjoying SSB and CW operating on VHF and UHF is knowing how to search for contacts. Unless there is a contest going on, you're unlikely to find a conversation the moment you flip the **POWER** switch. There's too much space and too few operators. To make the contacts, you have to work *smart*.

The first step is knowing how everyone else is operating, and to follow their lead. Essentially, this means to listen first. Pay attention to the segments of the band already in use, and follow the operating practices the experienced operators are using.

How are the Bands Organized?

In most areas of the country, everyone uses *calling frequencies* to establish contact. Then the two stations move up or down the band to chat. This way, everyone can share the calling frequency without having to listen to each other's conversations. You can easily tell if the band is open by monitoring the call signs of the stations making contact on the calling frequency. A complete list of calling frequencies is shown in Table 4-1.

SSB/CW activity on VHF/UHF is concentrated on the two lower VHF bands, 6 and 2 meters (50 and 144 MHz). The number of active stations on these bands is about equal. Above the 2-meter band, there are considerably more active stations on the 70-cm (432 MHz) band than any other.

On 6 meters, a *DX window* has been established to reduce interference to DX stations. Yes, *DX* stations! During years of high solar activity, 6-meter openings to the other side of the world are possible! Even during the "quiet" years 6 meters will occasionally open for contacts spanning 2000 miles or more.

The window, which extends from 50.100-50.125 MHz, is intended for DX contacts only. The DX calling frequency is 50.110 MHz. US and Canadian 6-meter operators should use the domestic calling frequency of 50.125 MHz for nonDX work. When contact is established, move off the calling frequency as quickly as possible.

Activity Nights

Although you can scare up a contact on 50 or 144 MHz almost any evening (especially during the summer), in some areas of the country there isn't always enough activity to make it easy to find someone. Therefore, informal *activity nights* have been established. There's a lot of variation in activity nights from place to place. Check with an active VHFer near you to find out about local activity nights.

Contacts *can* be made on non-activity nights as well. It may just take longer to get someone's attention.

Local VHF/UHF *nets* often meet during activity nights. (A *net* is a group of hams

Table 4-1
VHF/UHF Calling Frequencies

50.125 MHz
144.20 MHz
222.10 MHz
432.10 MHz

Common Activity Nights

Band (MHz)	Day	Local Time
50	Sunday	6:00 PM
144	Monday	7:00 PM
222	Tuesday	8:00 PM
432	Wednesday	9:00 PM
902	Friday	9:00 PM
1296 and up	Thursday	10:00 PM

who meet on the air to exchange information or discuss various topics.) Two national organizations, SMIRK (Six Meter International Radio Klub) and SWOT (Sidewinders on Two), run nets in many parts of the country. These nets provide a meeting place for active users of the 50 and 144-MHz bands. For information on the meeting times and frequencies of the nets run by SMIRK or SWOT, ask other occupants of the bands in your area, or see the Info Guide for more information.

What's that Beeping?

If you tune around on 6 meters when propagation conditions are good, you'll probably hear several *beacon* stations. Beacons send their call signs and other information in slow-speed Morse code. Other information may include their *grid-square* locations (see below), output power and antenna height. Most beacons use nondirectional antennas and relatively low power. If you can hear a beacon in a certain geographic area, you can probably work stations in that area. If you hear a beacon signal from several hundred or several thousand miles away, the band is open. Point your beam toward the beacon and start calling CQ! (Don't call CQ on the beacon frequency, though.) There are beacons on the bands above 50 MHz also, and they operate in a similar fashion. An extensive list of VHF beacons appears in *The ARRL Operating Manual*.

Grid Squares

One of the first things you'll notice when you tune the low end of any VHF band is that most conversations include an exchange of *grid squares*. Grid squares are a shorthand means of describing your general location anywhere on Earth. (For example, instead of trying to tell distant stations that, "I'm in Wallingford, Connecticut," I tell them, "I'm in grid square FN31." It sounds strange, but FN31 is a lot easier to locate on a map.)

Grid squares are coded with a 2-letter/2-number/2-letter code (such as FN24kp). This handy designator uniquely identifies the grid square and your exact location in latitude and longitude; no two have the same identifying code.

There are several ways to find out your own grid square identifier. The first bit of information you need is the *approximate* latitude and longitude of your station. (In most cases, the latitude and longitude of your city or town will be sufficient.) Your town engineer can provide this information, or you can go to a library and check a couple of geographic atlases. A nearby airport is another good source.

If you're fortunate enough to own a *GPS* (global positioning system) receiver, you can get precise latitude and longitude information in seconds. If you don't have

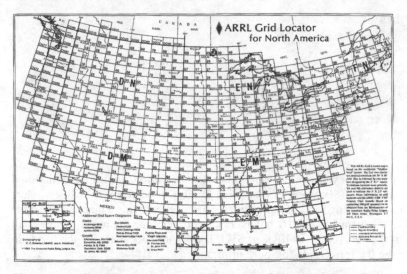

Figure 4-1—A grid-square map of the United States, similar to the one that appears in *The ARRL Operating Manual*. Grid-square designators are exchanged to qualify for contest points and various awards.

one of these wonders lying around, perhaps you know a friend who does.

With data in hand, you're ready to determine your grid square. If you have a copy of *The ARRL Operating Manual,* see Chapter 12. It includes a grid square map similar to the one shown in Figure 4-1. It also tells you how to convert your latitude and longitude to a grid square designator.

If you have a computer and a telephone modem, you can download GRIDLOC.ZIP from the ARRL BBS at 860-594-0306. This program for IBM-PCs and compatibles will instantly convert your latitude and longitude information to a grid square designator. You'll find similar software on CompuServe in the *HamNet* forum.

PROPAGATION

If you're new to the world above 50 MHz, you might wonder what sort of range is considered "normal." To a large extent, your range on VHF is determined by your location and the quality of your station. For example, a high-power station with a stack of beam antennas on a 100-foot tower will outperform a 10-W rig and a small antenna on the roof.

But for the sake of discussion, consider a more-or-less "typical" station. On 2-meter SSB, a hypothetical typical rig would be low-powered, perhaps a multimode transceiver (SSB/CW/FM), followed by a 100-W amplifier. The antenna of our typical station might be a single 15-element Yagi at around 50 feet, fed with low-loss coax.

Using SSB or CW, how much territory could this station cover on an average evening? Location plays a big role, but it's probably safe to say you could talk to similarly equipped stations about 200 miles away almost 100% of the time. Naturally, higher-powered stations with high antennas have a greater range, up to a practical maximum of about 350-400 miles in the Midwest (less in the hilly West and East).

On 222 MHz, a similar station might expect to cover about the same distance, and somewhat less (perhaps 150 miles) on 432 MHz. This assumes normal propagation conditions and a reasonably unobstructed horizon. This range is a lot greater than you would get for noise-free communication on FM. Increase the height of the antenna to 80 feet and the range might extend to 250 miles, and probably more, depending on your location. That's not bad for reliable communication!

Band Openings and DX

The main thrill of the VHF and UHF bands is the occasional band opening, when signals from far away are received as if they're next door! DX of well over 1000 miles on 6 meters is commonplace during the summer, and occurs at least a few times each year on 144, 222 and 432 MHz.

DX propagation on the VHF/UHF bands is strongly influenced by the seasons. Summer and fall are definitely the most active times although band openings occur at other times as well. Here is a review of the most popular types of VHF/UHF propagation. Remember that there is a lot of variation, and that no two band openings are alike. This uncertainty is part of what makes VHF/UHF interesting and fun!

- *Tropospheric—or simply "tropo"—openings.* Tropo is the most common form of DX-producing propagation on the bands above 144 MHz. It comes in several forms, depending on local and regional weather patterns. This is because it is caused by the weather. Tropo may cover only a few hundred miles, or it may include huge areas of the country at once. The best times of year for tropo propagation are from spring to fall, although they can occur anytime. One indicator of a possible tropo opening is dew on the grass in the evening. Another is a high-pressure weather system stalled over or near your location.
- *Meteor scatter* communication uses the ionized trails meteors leave as they pass through the atmosphere. VHF radio signals can be reflected by these high-altitude meteor trails and return to Earth hundreds or even thousands of miles away (see Figure 4-2). This ionization lasts only a second or so. Most meteor-scatter contacts are made on 6 and 2 meters. Because the ionization from a single meteor is brief, special operating techniques are used. (See the sidebar, "Hooked on Meteors.")

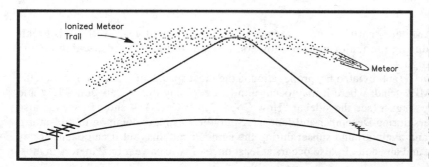

Figure 4-2—VHF radio signals can be reflected by the blazing trails of meteors as they enter our atmosphere. The results are quick contacts with stations hundreds or thousands of miles away.

Hooked On Meteors!

By Tom Hammond, WD8BKM

With three meteor contacts under my belt, I think I'm hooked!

I'd been active on 2-meter SSB for about three years, but I had never tried to make a meteor-scatter contact. The idea of bouncing my signal off the flaming tail of a meteor seemed absurd. Only the big-gun stations did that kind of stuff. My modest 2-meter setup (a Yaesu FT-736R, a small amplifier and a single Yagi) just didn't have enough *oomph*. Or did it?

Setting Up

It started Monday night, December 12, 1994, while I was listening to the Cincinnati-based VHF/UHF Net. I was hearing meteor bursts with astonishing frequency. The signals would pop out of nowhere for several seconds, then vanish utterly. If I could hear them, could I also *work* them? I jumped into the net and arranged a meteor schedule with AD4FF for the upcoming Geminids shower. We'd try to make contact Wednesday night on 144.143 MHz.

The next day, I posted electronic messages to the Internet *Usenet* groups rec.radio.amateur.misc and rec.radio.amateur.space requesting meteor skeds. I don't have HF equipment, so scheduling contacts on 75 meters (as many scatter buffs do) was not an option. I was happy to get a response from Rupert, N2OTO, on Long Island. We exchanged e-mail and set up a sked for 0330 UTC on December 14, following the sked with AD4FF.

Tuesday night, I listened on 144.200 MHz (the national SSB calling frequency) from 0230 UTC onward. The frequency was buzzing (or should I say "bursting"?) with activity. The meteor bursts were loud, but very short. I heard only partial call signs. I quickly decided that random meteor skeds should be left to the pros.

Timing is Everything

Timing is critical in meteor work. The typical procedure is to divide each minute into 15-second intervals. One station transmits for the first and third interval. The other station transmits for the second and fourth interval.

Each station calls the other giving both call signs (for example, "AD4FF WD8BKM"), repeating the calls over and over during the transmit interval. When you hear a burst with both call signs, you respond with a report indicating the length of the burn in seconds, such as "S2." The other station responds with, "Roger S2." The contact is complete when you exchange "rogers." I like to finish with, "Roger 73," if time permits.

Meteor scatter doesn't support much in the way of conversation, but it's a fascinating way to work distant stations on VHF or UHF. If you're chasing awards, you'll pile up the *grid*

Meteor-scatter contacts are possible at any time of year. Activity is greatest during the major meteor showers, especially the Perseids, which occurs in August.

• *Sporadic E* (abbreviated E_s) propagation is the most spectacular DX producer on the 50-MHz band, where it may occur almost every day during late June, July and early August (see the sidebar "How Good Can It Get?"). A short E_s season also occurs during December and January. Sporadic E is more common in mid-morning and again around sunset during the summer months, but it can occur at any time and any date. E_s also occurs at least once or twice a year on 2 meters in most *squares* quickly through meteor scatter.

First Contacts

I was all set to go at 0300 UTC. The first sked was easier than I thought. It took about 19 minutes to exchange calls, reports, and final "rogers" with AD4FF. I actually heard him just above the noise on two occasions, indicating some terrestrial propagation was going on as well. Bill is only about 450 miles from me, within range of most well-equipped stations under enhanced conditions.

I calmed my throat with ice-water, turned the antenna toward grid FN30 and relaxed in preparation for the next sked at 0330 UTC with N2OTO. By the start of the next sked, the heat sink on my amplifier had just cooled down. For the next 20 minutes, I called and called, and called, until I was blue in the face. I didn't hear a peep from Rupert. Perhaps east-west wasn't the best path for this shower.

I went back to the calling frequency and tried to pull together another contact. There were a lot of bursts, but most were too short to try a random contact. Just as I was about to pull the plug, I heard Gabor, VE3GBA, and decided to ask about his success. Gabor had already worked two skeds and was on his way to another with KB5IUA in grid EL29. From 0530 to 0600, I listened as Gabor and John exchanged reports. I heard John almost every minute of the sked, sometimes with sustained bursts of up to five seconds. It's surprising how long five seconds sounds during a meteor-scatter contact.

When they finished at 0600, I immediately started calling KB5IUA, hoping he might still be listening. I called for about three minutes before I finally decided to reach him by telephone. I have a directory on my computer that lists stations who are active on VHF and UHF. For each station the directory shows the grid locator, active bands and modes (EME, meteor scatter, etc) and an optional phone number to arrange schedules. John was very accommodating. I thought he wouldn't be too excited to work another station in grid EN82, but he responded, "I love working the rocks! Let's do it!"

We started our sked at 0608 UTC and were done within 10 minutes. On one of the last burns I heard John say, "73, Tom. You got it. That's a good one!" Twelve hundred miles on 2 meters via meteor. Amazing.

During our telephone conversation, John said he would be in grid EL18 in the morning at 1100 UTC. Could we try then? Why not?

With only 3$\frac{1}{2}$ hours of sleep, I was in front of the radio again. The 1100 sked took a little longer, with shorter bursts and weaker signals, but we made it. Instead of getting back in bed, I showered and headed off to work. I felt as if I had a hangover all day, but it was worth it!

areas. E_s results from small patches of ionization in the ionosphere's E layer. E_s signals are usually strong, but they may fade away without warning.

• *Aurora* (abbreviated Au) openings occur when the auroras are sufficiently ionized to reflect radio signals. Auroras are caused by the Earth intercepting a massive number of charged particles from the Sun. Earth's magnetic field funnels these particles into the polar regions. The charged particles often interact with the upper atmosphere enough to make the air glow. Then we can see a visual aurora. The particles also provide an irregular, moving curtain of ionization that can propagate signals for many hundreds of miles.

How Good Can It Get?

In 1995 hams were wallowing in the trough of a sunspot cycle. The days of almost daily DX on 6 meters seemed long gone, but the veterans knew better than to discount *sporadic E*. Sporadic-E propagation isn't affected by sunspots. Any time, any year, it can appear with surprising results. In the following excerpt from "The World Above 50 MHz" Emil Pocock, W3EP (September 1995 *QST*) chronicles a spectacular 6-meter sporadic-E event that had everyone buzzing for weeks!

Extraordinary Six-Meter Sporadic E!

What a month for sporadic E! Six meters opened to Europe and Africa on 15 days in June, for a total of nearly 60 hours. On two additional dates, North American stations worked the Azores, Madeira, and the Canaries. There have been European openings every June for the past half-dozen years or more, but never have there been so many, or the coverage so widespread. Although the Northeast often had the better of many of these events, stations scattered throughout the eastern half of the country were able to get into Europe.

Many 6-meter operators added new countries in their quest for DXCC—and even the old hands upped their country totals with the appearance of such sought-after stations as EH9IE (Ceuta and Melilla), EH6FB (Balearic Islands), IS0QDV (Sardinia), S59A and S57A (Slovenia), SP6RLA and other Poles, YO2IS and YO7VJ (Romania), and S0RASD (Western Sahara). In all, Americans and Canadians collectively logged more than 30 countries in Europe and North Africa.

As exciting as were the European openings, quite a stir was created from the Caribbean and Central America as well. On at least nine days, propagation favored the south from much of the US. The most sought after stations included TG9AJR (Guatemala), HP2CWB and HP3/KG6UH in Panama, and HR6/W6JKV (Honduras). US operators logged at least 16 different countries from the Caribbean region. Then there were the stations from the north! FP5EK (St. Pierre and Miquelon) provided a good deal of excitement both for Americans and Europeans on many days throughout the month. VE8HL (just south of the Arctic circle on Baffin Island) and OX3LX (Greenland) also worked stations throughout the eastern half of the US and Northern Europe.

Aurora-reflected signals have an unmistakable ghostly sound. CW signals sound hissy; SSB signals sound like a harsh whisper. FM signals refracted by an aurora are often unreadable. (Score another one for SSB and CW!)

• *EME, or Earth-Moon-Earth* (often called *Moonbounce*) is the ultimate VHF/UHF DX medium. Moonbouncers use the Moon as a reflector for their signals, and the contact distance is limited only by the diameter of the Earth (both stations must have line of sight to the Moon). As you've probably guessed, Moonbouncers have a particular obsession about knowing where the Moon is, especially when they can't see it because of cloud cover. Barking at the Moon indeed!

Moonbounce conversations between the USA and Europe or Japan are commonplace—at frequencies from 50 to 10,368 MHz. That's true DX! Hundreds of EME-capable stations are now active, some with gigantic antenna arrays. Their antenna systems make it possible for stations running 100 W and one or two Yagi antennas to work them. Activity is constantly increasing. In fact, the ARRL sponsors an EME contest, in which Moonbouncers compete on an international scale.

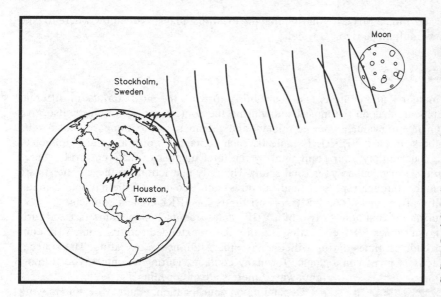

Figure 4-3—By reflecting signals off the surface of the Moon, a ham in Houston, Texas, can communicate with a ham in Stockholm, Sweden. This technique is known as *moonbounce*, or EME. If one station is using large antennas and high power, the other station can be very modest (100 W and a single beam antenna).

Hilltopping and Portable Operation

Maybe you'd like to try VHF/UHF weak-signal operating, but can't put up a tower. Or, you may live in a valley where hillsides block your signals. The solution? Take to the hills! VHF/UHF antennas are relatively small, and station equipment can be packed up and easily transported. Portable operation, commonly called *hilltopping* or *mountaintopping*, is a favorite activity for many amateurs. During VHF and UHF contests, a station located in a rare grid square is very popular. If you're on a hilltop or mountaintop, you'll have a very competitive signal.

Start by setting up on an easily accessible hill or mountain for an afternoon during a contest period. For a first effort, just take along a 2-meter rig. Even if you have an FM-only rig, you can still participate. Use the common simplex frequencies, like

EME (Moonbounce) with One 4-Element Yagi

By Emil Pocock, W3EP, from "The World Above 50 MHz," August 1995 QST

Do you need a huge station to operate Moonbounce? Not necessarily!

Chip Margelli, K7JA/6 (DM03), made a 2-meter EME contact with K5GW on May 5, 1995, at 0108 UTC using a single 4-element Yagi antenna! Chip's 900 W raised his effective radiated power to the same neighborhood as other moderately powered single-Yagi EME stations, but this must be the smallest antenna yet to make an EME contact. This was all the more remarkable because Chip could hear K5GW's signal on the small Yagi and then answered K5GW's CQ—no schedule was involved. Anyone want to try EME with a dipole?

146.52 and 146.55 MHz. If you find that the location "plays," you know where to take your new multimode rig next time!

VHF CONTESTING

Amateur Radio contests test your ability to work the most stations in different geographical areas on the most bands during the contest period. Contests also give you a chance to evaluate your equipment and antennas, and to compare your results with others. In most VHF/UHF contests, each contact is worth a certain number of points. You multiply your point total by the total number of different grid squares (*multipliers*) to calculate your final score. The only restrictions in these contests are that contacts through repeaters (and satellites) don't count, and the national 2-meter FM calling frequency, 146.52 MHz, is off limits for ARRL contest contacts.

During the first hour or two of a VHF contest, contacts may come fast and furious. At other times, VHF contesting is more like an extended activity hour. VHF contests provide set times during which many other stations are operating. The concentrated activity gives you a chance for many contacts. During a contest, you'll know right away if there's a band opening!

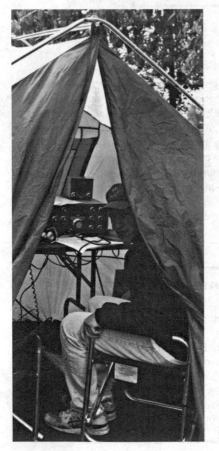

Depending on your location, you may be able to work dozens of different grid squares on several bands, which makes for a high score and lots of fun. If you're interested in awards chasing, you'll also be pleased to know that many hams travel to rare grid squares for contests.

Who Can Enter?

Most VHF/UHF contests are open to any licensed amateur who wants to participate. The ARRL sponsors all the major VHF/UHF contests (see Table 4-2), and specific rules, descriptions of the different categories, as well as entry forms, are available from ARRL Headquarters. You don't have to be an ARRL member to participate in these contests, nor are you required to submit your logs.

VHF/UHF contests feature a variety of categories among which you can choose. For single operators (those operating without assistance), entry classes in the ARRL contests include all-band, single-band, low-power portable, and one for Rovers (those who operate from more than one grid square during the contest). *The ARRL Operating Manual* is a good source of more information on selecting an entry category.

When and Why?

The ARRL VHF contests are held throughout the year,

VHF contesting doesn't always happen at home! Chris, N2OLW, takes a break from the WB2IEY/VE3 operation from Wolfe Island, Ontario.

Table 4-2

Major VHF/UHF Contests

(See QST magazine for complete details.)

Contest	Bands	When?
VHF Sweepstakes	50 MHz and up	Varies according to Super Bowl date.
Spring Sprints	One sprint per band	April/May
June VHF QSO Party	50 MHz and up	2nd full weekend
CQ Worldwide VHF	50 MHz and up	July
August UHF Contest	222 MHz and up	1st full weekend
September VHF QSO Party	50 MHz and up	2nd full weekend

with emphasis on the warmer months to encourage hilltop operation. (Who wants to freeze their toes off contesting from a mountain in sub-zero cold?) Outside of that, the ARRL VHF/UHF contest program is designed to take the best advantage of band openings that usually occur at certain times of the year. For instance, the June VHF QSO Party almost always occurs during periods of excellent sporadic-E propagation, giving you an opportunity to enjoy long-distance contacts on 6 and 2 meters. In fact, the first documented sporadic-E contact on the 222-MHz band was made during a June VHF QSO Party.

As shown in Table 4-2, the major ARRL VHF contests consist of the January Sweepstakes, June and September VHF QSO Parties, August UHF Contest, and the VHF/UHF Spring Sprints. Except for the Sprints, these events encompass many bands each. The January Sweepstakes and June and September QSO Parties are the most popular of them all, and each permit activity on SSB, CW and FM on all amateur frequencies from 50 MHz and up.

The UHF Contest is slightly different from the other contests described so far. The major difference is that only contacts on the 222-MHz and higher bands are allowed. The Spring Sprints are single-band, four-hour contests held over a several-week-long period. Sprints occur on the appropriate activity night for each band, so most Sprints are held on week nights. These short contests provide a super opportunity to test a new location or piece of equipment.

When to be Where

You'll find lots of random 6 and 2-meter activity during VHF contests. FM is relatively rare on 6 meters in the US, but it's quite common in most areas on 146, 222 and 440 MHz. On SSB, most stations stay near the calling frequencies of 50.125, 50.200, 144.200, 222.100 and 432.100 MHz. On CW, look between 80 and 100 kHz above 50, 144, 222 and 432 MHz. (Six meters offers less CW activity than the other VHF/UHF bands.) On FM, check the simplex calling frequencies (listed in Chapter 2), *except* 146.52 MHz.

BUILDING YOUR SSB/CW STATION

If you've read this far, I hope you're sold on the idea of trying SSB or CW on the VHF/UHF bands. Congratulations. You've reached the fun part, when you get

The ICOM IC-275A has been a popular multimode transceiver for several years. It offers FM, SSB and CW with a wide range of accessory functions.

The Kenwood TM-255A transceiver made its debut in 1995. In addition to being a multimode radio for SSB and CW work, it offers excellent 9600-baud packet performance.

An RF power amplifier like this one is a worthy addition to any serious VHF/UHF station. It takes the output of a multimode transceiver and boosts it to hundreds of watts. For meteor-scatter work, for example, a 150-W amplifier is considered to be the minimum requirement.

to open your checkbook (or grab your credit card) and start shopping for hardware. You have a lot of options available, depending on how much you want to spend.

What's Out There?

Multimode VHF transceivers can be grouped into two classes: home station and mobile/portable. A look at *The ARRL Radio Buyer's Source-book* will help you decide what's right for you. It's also a good idea to review recent *QST* Product Reviews when you're selecting equipment.

Many people just getting into VHF settle on multimode, single-band mobile or portable transceivers. These rigs are often less expensive, less complex and more flexible (in terms of power sources and size) than home-station rigs. Some home-station rigs include accessories not usually found in portable and mobile rigs. Serious operators find these accessories, such as preamplifiers, narrow IF filtering and noise blankers helpful when propagation and interference conditions make it hard to hear another station.

Although most VHF multimode transceivers are single-band radios, multi-

How Much Will It Cost?

There are many ways to set up a VHF/UHF station, depending on what you want to do. Here are some equipment lists based on the discussion in this chapter. The lists are broken down by activity.

Note that I'm basing these estimates on *new* equipment. If you shop around for used gear, you can save as much as 50% or more.

Basic SSB/CW Station on 50, 144, 222 or 430 MHz

Multimode transceiver:	$1000
Power supply (to run the transceiver):	$200
11-element beam antenna:	$200
Rotator:	$100

Add Meteor-Scatter/Aurora Capability

150-W amplifier:	$350

Add 144-MHz Moonbounce Capability

High-current power supply:	$400
Four 11-element beam antennas "stacked" together:	$800
Heavy-duty rotator:	$250
300-W amplifier:	$600

band transceivers have been growing in popularity. Usually aimed at the amateur satellite market, these rigs are also popular among terrestrial operators because of their flexibility. They usually allow you to receive on one band while transmitting on another. These rigs are considerably more expensive than their single-band counterparts, but less expensive than buying separate radios for each band they cover.

Transverters

An alternative to buying one or more VHF transceivers is to buy or build a *transverter* to accompany your HF rig. A transverter takes the RF output from your HF transceiver and uses it to create a signal on a particular VHF or UHF band. The transverter also converts received VHF and UHF signals to HF frequencies. In effect, a transverter turns your HF transceiver into a VHF or UHF transceiver.

Although this equipment requires some effort to interface with an HF rig (except for those made to go with your particular transceiver), the performance and cost savings can be substantial.

And if you don't own an HF transceiver, consider buying a used rig as a "platform" for your transverter. During the solar cycle peak that occurred in the late '80s and early '90s, several manufacturers sold inexpensive 10-meter transceivers. These little rigs were great for working the world with relatively low power (25 W) when the 10-meter band was hot. When the band "cooled" in the mid-'90s, hams started selling these radios at bargain prices. Wise amateurs snapped them up and they soon became the hearts of many VHF and UHF stations!

There are several transverter manufacturers who market to US hams, including . . .

Down East Microwave
954 Rte 519
Frenchtown, NJ 08825
tel 908-996-3584

SSB Electronic
124 Cherrywood Dr
Mountaintop, PA 18707
tel 717-868-5643

Hamtronics
65-Q Moul Rd
Hilton, NY 14468
tel 716-392-9430

Advanced Radio Technology
Suite 2 Spence Mills
Mill Lane, Bramley
Leeds LS13 3HE
England
tel 011-44-113-236-9973

VHF/UHF Antennas

We had a pretty robust discussion of VHF and UHF antennas in Chapter 2. The same principles apply with one exception: antennas for SSB and CW operating must be *horizontally polarized*. That is, the parts of the antennas that radiate and receive energy must be *parallel* to the ground.

Why?

Because that's the way everyone else does it! Seriously, it's more than a matter of group consensus. If your antenna is vertically polarized and the other fellow is using a horizontally polarized antenna, his signal will be substantially weaker than it should be on your end, and vice versa.

The loss caused by mismatched polarization is most noticeable during line-of-sight contacts. If your signal is bouncing off a meteor trail or a layer of the atmosphere, polarization doesn't mean much. What goes up vertical, for example, comes down every which way!

Because we're dealing with "weaker" signals, directional antennas are best. You want to focus your signal as much as possible. There is no law that says you can't use an omnidirectional antenna on SSB or CW, but your effective range will be very limited.

Directional antennas include the venerable Yagi and the quad. The Yagi is the most common directive antenna. Yagi antennas are commercially available with three to at least 33 elements. The quad uses loop elements instead of wire or rod elements.

If you want multiband performance in a single directional antenna, consider the *log-periodic dipole array* (LPDA), usually referred to simply as a *log periodic*. A log periodic beam antenna covers several bands. The penalty for this frequency coverage is significantly lower gain than can be achieved with a single-band Yagi or quad. Log periodics are also more mechanically complex than Yagis and quads. On the other hand, the convenience of having coverage of so many bands with only one antenna and feed line is very attractive, especially for portable operation.

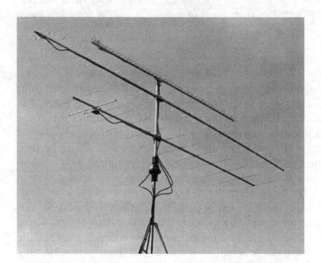

Beam antennas like these are *essential* for SSB and CW operating on the VHF and UHF bands. If you attempt to use an omnidirectional antenna, you'll discover that your range is severely limited. Beams focus your power in a particular direction, so you'll need an antenna rotator to turn them. If you use small beams, a simple TV-type rotator is adequate for the task.

When you *really* need to focus your signal, you can "stack" several beams together so that they function as one antenna! This type of antenna is often used by the big VHF contest stations and by serious moonbouncers. Stacking is an expensive option and a bit of an art form. Make sure your bank account is solvent and that you enlist the aid of a veteran ham before you attempt an antenna system like this!

How Do I Choose?

There's even more variety in antennas than in equipment for the VHF bands. Unless your local dealer carries all the major brands, you'll do most of your shopping without actually seeing all available antennas up close. That's why it's a good idea to get catalogs from major retailers who advertise in *QST*, which show at least the major specifications for the antennas they carry. Also, read *QST* Product Reviews, and ask your friends what they're using.

Feed Lines: The Weak Link

When you install your antennas, you'll need to connect them to your radios via feed lines. No surprise so far, right? What makes this subject worth discussing here is *loss*.

If you missed the discussion of feed line loss in Chapter 2, go back and read it. Choosing the right coaxial cable is important for FM operating . . . and it's downright *critical* for SSB and CW! Consult the table in Chapter 2 and decide conservatively. That is, always choose the lowest-loss cable you can afford.

Don't blow your savings on a nice transceiver and a whiz-bang antenna only to scrimp on the coax. You'll be very sorry. Many hams have tried their hands at "weak signal" operating, but they were handicapped from the start because they bought coax that was way too lossy for the installation. Not only did they lose precious transmit energy, their received signals were watered down as well.

THE CHALLENGE AND THE REWARD

There's no question that it's easier to get on the air with FM than SSB or CW. With FM, it may be a matter of simply buying a hand-held transceiver and talking through your local repeater. SSB and CW take a little more effort, but the reward is considerable!

As a "weak-signal" operator, you'll enjoy contacts over distances that FM enthusiasts can only achieve through complex linked-repeater systems. Best of all, you'll experience the true magic of VHF operating. As you sharpen your skills, you'll be able to predict when band openings are about to take place. By listening to the distant signals, you'll know which propagation mode is active and how to use it to your advantage.

VHF operating will challenge you every day. DX stations sometimes appear when you least expect them—and disappear just as suddenly. Wait until the day when you turn on your equipment and hear a flood of distant CW and SSB signals. The excitement will be electrifying and you'll know in that moment what you've guessed all along: there is much more to VHF than FM!

Outta Space!

Ground control to Major Tom.
Ground control to Major Tom.
Take your protein pills
And put your helmet on . . .

— David Bowie, *Space Oddity*

It's a long way from Starfleet, but there is an armada of Amateur Radio space-craft in orbit above our planet at this very moment. (Quick, ma! Grab the telescope!) When this book went to press, "Hamfleet" was comprised of 15 satellites. By the end of the decade, that number will probably grow to 25.

Believe it or not, amateur satellites have been in orbit since the early '60s. Even before astronaut John Glenn made his historic flight, OSCAR 1 (*O*rbiting *S*atellite *C*arrying *A*mateur *R*adio) was circling the Earth, transmitting "HI" in CW.

Today you can choose from a variety of extremely sophisticated amateur satel-lites. You can even communicate with Russian cosmonauts and American space shuttle astronauts. What may surprise you more than anything else, however, is the ease with which you can access most of these satellites. It seems to be one of the best-kept secrets in Amateur Radio!

FINDING THE SATELLITES

Before you can use any amateur satellite, you need to find it—or at least be able to predict where it will be at a given time. This is a little more complicated than it sounds. Now don't get intimidated! "Complicated" doesn't mean "difficult." It just takes a few extra steps. Allow me to explain . . .

As of this date, all Amateur Radio satellites are in nongeostationary orbits. This simply means that the satellites are not in fixed positions in the sky from our perspective here on Earth. They are like tiny moons, rising and setting over your local horizon.

If a low-altitude satellite passes over your house at 9 AM, it will make a complete orbit and return to the same spot about 100 minutes later . . . or will it? It might, *if the Earth remained perfectly still*. But the Earth is turning at a pretty good clip and it takes you, your house, your car and everything else in your neighborhood right along with it. By the time the satellite completes its orbit, you'll be several hundred miles east of where you were before. So, the satellite won't appear over your house again; it will be screaming over someone else's home.

The problem gets stickier when you talk about satellites that have unusual orbits. For example, some satellites zoom in very close to the Earth, then go flying far out into space. (Imagine a kind of cosmic yo-yo.)

Orbital Elements

How can you know when a satellite is about to make an appearance in your neighborhood? To answer that question you need to know the satellite's *orbital elements*.

An orbital element set is merely a collection of numbers that describes the movement of an object in space (see Figure 5-1). By feeding the numbers to a computer program, you

Figure 5-1—Looks intimidating, doesn't it? This is an orbital-element (a.k.a. "Keplerian") set for the RS-15 satellite as displayed by the *Instantrack* program. If you want to learn the meanings of all these numbers and abbreviations, more power to you! You'll greatly expand your understanding of orbital mechanics by doing so. On the other hand, if you just want to work an amateur satellite, all you need to do is feed this data to your satellite-tracking program. Some programs, such as *Instantrack*, will load the data automatically if you provide it in the form of an ASCII file. Other programs require you to enter the data manually.

can determine exactly where a satellite is (or will be) at any time. Before the heyday of PCs, hams did their orbital calculations manually. You can learn a lot about orbital mechanics from the manual method, but few prefer to do it that way these days.

Finding the Elements

There are several sources for orbital elements:
- Satellite newsletters (see the Info Guide)
- W1AW RTTY and AMTOR bulletins
- Packet bulletin boards
- ARRL telephone BBS (860-594-0306)
- CompuServe (*HamNet* forum)
- The Internet Worldwide Web (http://www.qualcomm.com/amsat/AmsatHome.html)

If you have an HF radio, RTTY capability, a packet TNC, a telephone modem or the necessary cash for a subscription, you'll always be able to get the latest orbital elements for the satellites you want to track. If all else fails, there is probably someone in your area who has access to the elements. Ask around at your next club meeting. With a couple of exceptions, you only need to update your elements every few months.

Feeding the Software

There are many *satellite-tracking* programs on the market that will take your

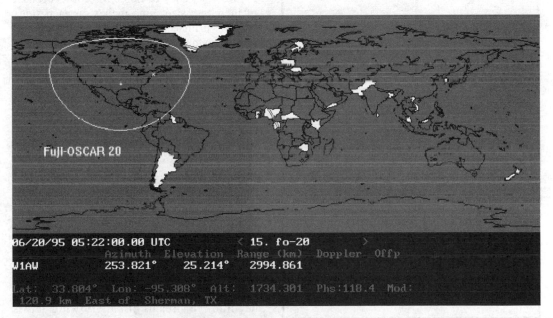

Figure 5-2—This is one of two tracking screens provided by the *Instantrack* satellite-tracking program. In this case, we're looking at the position of Fuji-OSCAR 20 over North America. Most stations located within the white circle (known as a *footprint*) should be able to talk to each other through the satellite. The program also shows the satellite's position relative to your station (azimuth, elevation, range). Most satellite-tracking programs offer similar displays.

orbital elements and magically produce satellite schedules. Some will read the elements indirectly; others require you to type them in manually.

Among other things, these programs tell you when satellites will appear above your local horizon and how high they will rise in the sky (their elevation). When working satellites, the higher the elevation the better. Higher elevation means less distance between you and the satellite with less signal loss from atmospheric absorption.

Some programs also display detailed maps (Figure 5-2) showing the ground track (the satellite's path over the ground). AMSAT (the Radio Amateur Satellite Corporation) offers satellite tracking software for a variety of computers. For more information contact: AMSAT, PO Box 27, Washington, DC 20044, tel 301-589-6062.

Let's assume that you have a satellite tracking program and you've provided it with the latest orbital elements. You're all set to track satellites, so let's start with an easy one . . .

ON THE WINGS OF A DOVE

Do you own a 2-meter FM rig? How about an outdoor antenna of some kind? Excellent! You are now the proud owner of a basic satellite receiving system—and there is a bird in orbit just waiting to "talk" to you. It's called DO-17, otherwise known as *DOVE*.

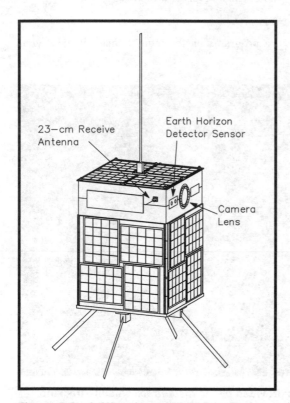

23—cm Receive Antenna

Earth Horizon Detector Sensor

Camera Lens

Figure 5-3—A *Microsat* is little more than a tiny cube (9 inches square) stuffed with electronics and covered with solar cells. This drawing depicts WEBERSAT-OSCAR 18.

DOVE is one of several *Microsats* (see Figure 5-3). They're called Microsats because of their tiny size (nine inches on each side). DOVE's primary mission is education. It transmits streams of packet telemetry and occasional voice bulletins on 145.825 MHz. By studying the telemetry, you can learn all sorts of fascinating things about conditions in space. Since DOVE is a *LEO* (*Low Earth Orbiting*) satellite, its signal is very easy to hear. (You can even receive DOVE on so-called "police scanners.")

If you only want to listen, you'll get an earful of raucous packet bursts as it streaks overhead. DOVE also has digital voice capability and may be transmitting in that mode from time to time.

If you have packet equipment, you're in for an extra treat. Set up your packet TNC as you would for normal operation and switch your FM transceiver to 145.825 MHz. As DOVE rises above the horizon you'll begin to see streams of data flowing across your monitor. You may also see brief text bulletins.

After you get tired of watching raw data, you'll want to find out what it means. There

```
uptime is 141/09:16:30. Time is Mon Mar 28 03:29:46 1994
```

DOVE		Array V	9.941 V	BCR Load Cur	0.080 A
+10V Bus	11.147 V	BCR Set Point	20.213 C	X Array Cur	0.011 A
+8.5V Bus Cur	0.029 A	+5V Bus Cur	0.229 A	+Y Array Cur	0.011 A
+X Array Cur	0.013 A	Y Array Cur	0.010 A	Ext Power Cur	0.020 A
–Z Array Cur	0.017 A	+Z Array Cur	0.011 A	Bat 1 Temp	4.234 D
BCR Input Cur	0.188 A	BCR Output Cur	0.017 A	FM TX#1 RF OUT	0.026 W
Bat 2 Temp	–24.206 D	Baseplt Temp	3.024 D	+Y Array Temp	–0.607 D
FM TX#2 RF OUT	0.003 W	PSK TX HPA Tmp	–18.760 D	+Z Array Temp	–10.288 D
RC PSK HPA Tmp	–5.448 D	RC PSK BP Temp	–3.027 D	Mixer Bias V	1.367 V
Rx E/F AudioW	2.165 V	Rx E/F AudioN	2.165 V	Rx A Audio (N)	2.165 V
Osc. Bisd V	0.500 V	Rx A Audio (W)	2.165 V	Rx E/F DISC	–1.080 k
Rx A DISC	0.318 K	Rx A S meter	83.000 C	+5V Rx Current	0.023 A
Rx E/F S meter	112.000 C	+5 Volt Bus	4.880 V	IR Detector	0.0000 C
+2.5V VREF	2.506 V	8.5V BUS	8.367 V	GASFET Bias I	0.004 A
LO Monitor I	0.001 A	+10V Bus	10.454 V	+X (RX) temp	–4.237 D
Ground REF	0.000 V	+Z Array V	0.205 V	Bat 2 V	1.289 V
Rx Temp	6.654 D	Bat 1 V	1.289 V	Bat 5 V	1.289 V
Bat 3 V	1.290 V	Bat 4 V	1.280 V	Bat 8 V	1.303 V
Bat 6 V	1.319 V	Bat 7 V	1.298 V	+8.5 V Bus	8.344 V
		+5V Bus	4.802 V		

A sampling of packet telemetry from the DOVE (OSCAR 17) satellite.

are several programs available to decode DOVE telemetry. For more information, send a self-addressed, stamped envelope to AMSAT at the address shown above. Ask for their software catalog.

DOVE has had a troubled history, with several failures during its career. Each time, however, AMSAT volunteers have managed to revive the satellite and get it back into working order. When DOVE is operating, it pumps out a strong signal. I've heard it clearly on a hand-held transceiver with just a rubber duck antenna.

THE *MIR* SPACE STATION

You've discovered how easy it can be to eavesdrop on satellite signals. Now it's time to start thinking in terms of *transmitting* to a satellite. A perfect candidate for your first transmission is the Russian *Mir* space station.

Mir has been occupied by Russian cosmonauts for several years as a laboratory for testing human responses to long-duration space flights. The *Mir* studies are extremely important for future manned missions to Mars and beyond.

To combat boredom, an Amateur Radio station was installed. The cosmonauts pass amateur license tests and are assigned special *Mir* call signs (such as RØMIR) prior to launch. When they reach the station, they operate 2-meter FM voice or packet.

Like the DOVE satellite, *Mir*'s signal is powerful. You'll usually find it on 145.55 MHz, and you won't need sophisticated equipment to hear it—or to be heard. Once

The Russian *Mir* space station orbits 300 miles above the Earth. The cosmonauts use FM voice and packet radio to keep in touch with hams.

again, an outside antenna—such as a ground plane—works fine. Its orbit provides a couple of very good "passes" each day for most areas.

Mir on Packet

The *Mir* Amateur Radio station uses standard 1200-baud packet—the same packet format you use here on Earth (see Chapter 3). The *Mir* packet station includes a mailbox where you can leave messages for the cosmonauts (or anyone else) and pick up their replies.

The biggest problem with working *Mir* on packet is interference—lots of interference! With the signal coverage the space station enjoys, you can imagine how many hams might be trying to connect to *Mir* at the same time. This creates pure chaos as far as its FM receiver is concerned.

If you're able to connect to the mailbox, the constant bombardment of signals may make it difficult for you to post your message. (Remember that you may only have a few minutes before the space station slips below your horizon.) Here are a couple of tips to improve your chances:

- Listen before you start sending your connect requests. Monitor a few transmissions and make sure you have the correct call sign. The call sign changes whenever a new crew occupies the station.
- Use as much power as you have available. If there were only a couple of stations competing for *Mir*'s receiver, you'd only need a couple of watts to have a decent chance of connecting. During a normal pass, however, there are usually dozens of stations blasting out connect requests. The stations that pack the bigger punches seem to win consistently.
- Try connecting during "unpopular" hours. If you have the stamina to sit up and wait for a late-night pass, you may have a better opportunity to make a connection.

When you finally connect to the mailbox, make your message entry short. The station will be out of range before you know it and other hams will be waiting to try their luck. Some packet software permits the user to create a message file before attempting to connect. If your software offers this feature, it will come in handy for *Mir*.

Voice Contacts with *Mir*

The *Mir* cosmonauts obviously enjoy packet, but sometimes they crave the sounds of other human voices. You may be waiting for a chance to connect on packet, only to hear them calling CQ instead!

Working *Mir* on voice is very similar to working a DX pileup. You sit with microphone in hand and wait until you hear the cosmonaut complete an exchange. At that moment you key the mike and say your call sign. Now listen. No response? Call again quickly! Keep trying until you hear him calling you or someone else.

I've heard of hams working *Mir* while mobile and some claim to have worked *Mir* with hand-helds. As you might imagine, *Mir* QSL cards are highly prized!

The major problem with working *Mir*—on voice or packet—is its erratic schedule. The cosmonauts have many daily assignments and are not always able to find the time to operate their amateur station. They are sometimes forced to turn off their equipment altogether to avoid interference to other systems during critical tests.

Another problem concerns *Mir*'s orbit. The space station travels at a relatively low altitude, so it's always subject to a significant amount of atmospheric drag. If it didn't occasionally "boost" to a higher orbit, the station would reenter the atmosphere and be destroyed. Every time *Mir* fires its rocket engines to adjust its orbit, a revised set of orbital elements must be distributed. If you want to try your luck with *Mir*, plan to update your elements for the space station as often as possible.

SAREX

SAREX, the Shuttle Amateur Radio Experiment, is a continuing series of Amateur Radio operations from US space shuttle missions. The first SAREX operations employed 2-meter FM voice, but more recent flights have also used packet. The vari-

Astronaut Linda Godwin, N5RAX, pauses during her mission duties to operate the SAREX station aboard the space shuttle *Endeavor*. (*Photo courtesy of NASA*)

ety of modes in use depends on the available cargo space. In addition, not every shuttle astronaut is a licensed ham, so not every shuttle mission has an active SAREX operation. Check *QST* for the latest news on upcoming SAREX missions.

Unlike *Mir*, SAREX uses a split-frequency uplink/downlink scheme. Earthbound DXpeditions use split-frequency operation to maximize the number of stations they can work. The same is true for SAREX. (If operating from outer space isn't a DXpedition, I don't know what is!) With this thought in mind, you can appreciate the importance of knowing which frequencies are being used for the uplinks and downlinks. (They will be published in *QST* well in advance of the launch date.) Whatever you do, never transmit on the SAREX downlink frequency. This mistake will make you very unpopular very quickly!

You'll need the shuttle's orbital elements to predict when it will be in range and these will be available through the sources we've already discussed. You can use the same 2-meter equipment for SAREX as you do for *Mir*. During one of the "packet robot" operations on a previous flight, I managed to connect using my trusty ground plane.

THE RS (RADIO SPUTNIK) SATELLITES

Have your ever heard of RS 10, 12 or 15? They are unmanned Amateur Radio satellites placed in orbit by the former Soviet Union. Without a doubt they are among the easiest satellites to work.

The RS satellites are completely different from DOVE, *Mir* or the space shuttle. They are basically orbiting SSB/CW *repeaters* riding piggyback on larger satellite platforms. All RS satellites are equipped with unique devices called *linear transponders*.

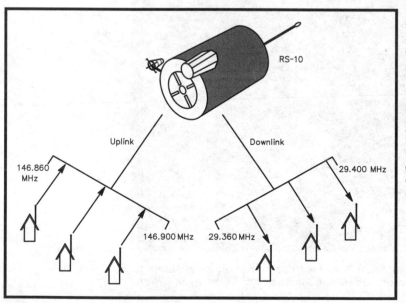

Figure 5-4—A linear transponder acts much like a repeater, except that it relays an entire group of signals, not just one signal. This is an illustration of the linear transponder aboard RS-10. It listens on the 2-meter band and repeats everything it hears on a portion of 10 meters.

Table 5-1

Amateur Satellites: Frequencies and Modes

Satellite	Uplink (MHz)	Downlink (MHz)
SSB/CW		
AMSAT-OSCAR 10	435.027—435.179	145.825—145.977
AMSAT-OSCAR 13	435.423—435.573	145.825—145.975
	435.601—435.637	2400.711—2400.747
Fuji-OSCAR 20	145.900—146.000	435.800—435.900
RS-10	145.860—145.900	29.360—29.400
RS-12	21.210—21.250	29.410—29.450
RS-15	145.858—145.898	29.354—29.394
Packet—1200 bit/s		
(FM FSK uplink, PSK downlink except as noted)		
AMSAT-OSCAR 16	145.90, .92, .94, .96	437.05/437.026
DOVE-OSCAR 17	None	145.825
Telemetry only. FM FSK downlink.		
WEBERSAT-OSCAR 18	None	437.10
Telemetry and images only.		
LUSAT-OSCAR 19	145.84, .86, .88, .90	437.126/437.15
ITAMSAT-OSCAR 26	145.875, .900, .925, .950	435.870
Mir Space Station	145.55	145.55
Packet mailbox. FM FSK simplex.		
Packet—9600 bit/s		
(FM FSK uplink and downlink.)		
UoSAT-OSCAR 22	145.900, .975	435.120
KITSAT-OSCAR 23	145.85, .90	435.175
KITSAT-OSCAR 25	145.87, .98	436.50
FM Voice		
AMRAD-OSCAR 27	145.850	436.800
Repeater.		
Mir Space Station	145.55	145.55
Occasional simplex QSOs with the cosmonauts.		

Linear Transponders

Earthbound repeaters listen on one frequency and repeat what they hear on another. Imagine what would happen if your local repeater could retransmit everything it heard on an entire group of frequencies—not just one conversation, but *several at once*? This is exactly the function of a linear transponder.

The transponders aboard RS-10 and RS-15 listen to a portion of the 2-meter band and retransmit everything they hear on the 10-meter band (see Figure 5-4). This type of operation is known as *Mode A*. The transponder aboard RS-12 listens to a section of the 15-meter band and simultaneously retransmits on 10 meters. This is known as *Mode K*. The range between the highest and lowest uplink (or downlink) frequencies is known as the transponder's *passband*. See the RS uplink and downlink passbands in Table 5-1.

Not only do linear transponders repeat everything they hear on their uplink pass-bands, they do so very faithfully. CW is retransmitted as CW; SSB as SSB. FM voice transmissions are strongly discouraged since their broad signals occupy an enormous chunk of the downlink passband. Not only would this limit the number of stations that could use the satellite, it would place a severe drain on transponder power.

Working the RS Satellites

For RS-12 all you need is an HF transceiver capable of split-frequency operation on separate bands (15 and 10 meters). For RS-10 and RS-15 you'll need a 2-meter multimode (SSB/CW/FM) transceiver and a 10-meter SSB receiver (see Figure 5-5).

There is one "legal" catch with RS-12. Take another look at the RS-12 uplink passband in Table 5-1. If you have a chart of US amateur HF frequencies, you'll quickly discover that you must have an Advanced or Extra Class license to transmit within RS-12's uplink passband. (Now there's a decent incentive to upgrade your license!)

All hams except Novices can use RS-10 and RS-15. If you're a Technician, you might think that you're out in the cold, too. It's true that Technicians can't legally operate within the 29-MHz downlink passbands these satellites utilize, but are you actually *operating* there? Think about it. The satellites are repeating your signals at 29 MHz, but *they're* doing the transmitting, *not you*. You're transmitting to the satellite on 2 meters, and that's perfectly legal. The bottom line is: Don't let anyone tell you that you can't use RS-10 or RS-15 because you're a Technician. Instead, politely advise them to reread their *FCC Rule Book*!

No Multimode Equipment?

"I can receive on 10 meters, but I only have a 2-meter FM rig. I'll never be able to work RS-10 or RS-15," the skeptic grumbles. On the contrary! Have you considered

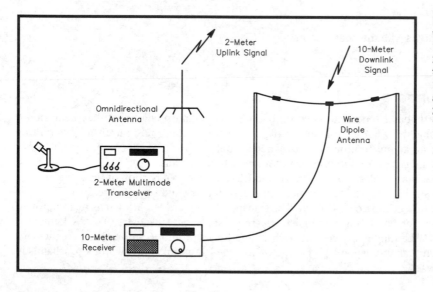

Figure 5-5—To work RS-10 or RS-15, you'll need a 2-meter multimode transceiver, a 10-meter receiver and antennas for 2 meters and 10 meters.

Finding Your Downlink Signal on RS-10

If you know your uplink frequency, how can you predict where your signal will appear in the RS-10 downlink passband? There is a calculation you can use that will get you in the ballpark:

$$F_{Down} = F_T + F_{Up}$$

F_T is the *translation constant* in MHz and F_{Up} is your chosen uplink frequency. For RS-10 the F_T value is −116.5 MHz.

Let's assume that you've chosen 145.870 MHz as your uplink frequency.

−116.5 + 145.87 = 29.37 MHz

See how easy it is? If you're transmitting on 145.87 MHz, you can expect to find your downlink signal in the vicinity of 29.37 MHz.

---

using your FM transceiver as a CW rig? All you have to do is wire a key to the push-to-talk (PTT) pins on a spare microphone plug. Your signal may be a little raw and "chirpy," but you'll be transmitting usable CW!

Antennas

Elaborate antennas are definitely not required to work the RS satellites. A wire dipole is fine for receiving the 10-meter downlink signal. By the same token, a basic ground plane is adequate for your 2-meter uplink to RS-10 or RS-15. A 15-meter dipole is fine for an uplink to RS-12. In terms of power, 20 to 30 W seems to work well—although I managed to work a station through RS-10 with only 5 W. For RS-12, most hams are using their standard HF transceivers and running about 100 W.

With the wide separation between uplink and downlink frequencies, you can work the RS satellites in *full duplex*. (With RS-12, however, you'd need a separate 10-meter receiver for full-duplex work.) When operating full duplex, you can hear your own signal on the satellite downlink *while you're transmitting*! When you're listening to your own signal through the satellite, you'll notice something odd right away—your signal keeps drifting downward in frequency. That's why you need to tweak the VFO on your uplink transmitter to keep your signal on frequency. All these acrobatics are caused by a pesky problem known as *Doppler shift*.

The Mystery of the Shifting Signal

Doppler shift is caused by the difference in relative motion between you and another object—a satellite in this case. As the satellite moves toward you, the signal frequencies in the downlink passband gradually *decrease*. It's the same effect you hear when a locomotive approaches a crossing. The audio frequency of the horn decreases as the train moves toward you, then away. The higher the frequency, the more pronounced the effect. Doppler, for example, is much more noticeable on 435-MHz signals than on 29-MHz signals.

Dealing with Doppler is not that difficult once you get used to it. The standard rule of thumb is to change the frequency of your *uplink* to compensate, while leaving your downlink frequency the same. For example, when you transmit SSB through RS-10 on 145.870 MHz, you hear your own signal being repeated at about 29.370

RS-12 Worked All States

By R. A. Peschka, K7QXG

There was a time when I thought satellite operations were reserved for the technically elite of our Amateur Radio society. That misguided perception prevailed until April 1993 when I overheard an interesting conversation between Roger, N4ZC, and a station in Puerto Rico. N4ZC was encouraging the KP4 to try a contact on the next pass of the RS-12 satellite using mode K (mode K means you transmit on 21 MHz and listen on 29 MHz). Their conversation intrigued me so much I decided to give it a try myself. A short time later I had made contact with N4ZC via the RS-12 satellite and my space odyssey began.

Sharpening My Skills for WAS

N4ZC proved to be a great help. He sent me a list of the times when I could expect the "bird" to pass within my range. It wasn't long before it became clear that the use of a good computer program would go a long way toward making these operations even more fun and rewarding. I joined AMSAT; acquired a good program for my PC; and, now charged with excitement, was active on the satellite. After a month or so, it became apparent that the challenge chasing my Worked All States

(WAS) award on RS-12 would be an excellent way to tune up my operating skills, which had become quite rusty.

At first I evaluated each projected pass of the satellite. I designed a precise strategy; but I threw all my good intentions to the wind when the thrill of the hunt took over. I'm sure I missed many golden opportunities to contact some of those hard-to-reach stations.

I began making progress when I took my methods seriously. After every pass I evaluated the results to see what worked—and what didn't. Part of the trick was becoming intimately familiar with every little knob and dial on my transceiver. I began to spend far more time listening than transmitting, and learned how to compensate for noise, Doppler shift, and all the other nuances encountered when working the "bird." I honed my skills with practice and the states fell one by one.

Down to the Last States

After nearly a year, I needed only five states to reach my goal: South Dakota, Wyoming, Montana, Delaware and Maine. A good friend and fellow satellite hunter suggested

MHz. As you listen, you hear your signal shifting downward. Don't retune your receiver! Instead, tweak your *transmitter* VFO as much as it takes to keep your voice sounding normal. We'll discuss this technique in more detail later.

The "Flavors" of the RS Birds

Each of the RS satellites has a particular "flavor" that's dictated by its operating mode, orbit and so on. The RS bird you'll choose to operate will depend on what you want to accomplish.

RS-10 is the workhorse for medium-range contacts. Its range spans roughly 2000 miles, which makes it a favorite among hams who are chasing their Worked All States and VHF/UHF Century Club awards. Contacts on this satellite tend to be short. Activity can be heavy on weekends.

RS-15 orbits nearly twice as high as RS-10, but its signal output is 50% less. With its higher orbit you can enjoy much longer conversations through RS-15. However, RS-15 has a meager power budget and a small solar-cell capture area. This means that operation

Clarence, N7RPC in Wyoming. I fired off a letter to Clarence, seeking a schedule on RS-12. The letter resulted in a telephone call, a bit of preparation and a QSO attempt. After some momentary confusion, followed by another quick telephone call, we made contact. Meanwhile, by pure luck, I managed to find W0IT in South Dakota.

Clarence, in turn, suggested Ken, WG7G, in Montana as a possibility to move up one more notch. A letter was sent to Ken, but before his reply arrived he found *me* on the satellite! With Montana in the bag I had a total of 48 states! Only Delaware and Maine remained elusive.

As I chased the states I renewed operating skills long forgotten, not the least of which was patience and courtesy. I once again became familiar with the nuances of having pens run out of ink at the precisely wrong moment; of having two too many thumbs, and of drinking too much coffee before the "window" opened to the usual flurry of activity.

The search for Delaware and Maine continued. Finally, I heard Gene, NY3C, in Delaware, but the satellite sank below the horizon before I could make contact. It was letter writing time again. Gene really had the true ham spirit. He called me by telephone and we arranged a list of possible schedules. We made contact on the very first try. That left only Maine.

I had sent letters to a couple of stations in Maine, but never received a response. NY3C stepped into the breach again and spread the word of my quest on various nets. Two days later, on my favorite CW frequency, I was called by W1OO in Maine!

About RS-12

This amateur satellite is actually part of a large Russian COSMOS navigational satellite launched in 1988. It was assembled at the Tsiolkovskiy Museum for the History of Cosmonautics in Kaluga (about 180 km southwest of Moscow).

Although it is capable of operating in a number of modes, RS-12 only functions in mode K at this time. The frequencies are shown in Table 5-1.

Elaborate antennas and high output power are *not* required to work RS-12. Simple dipole antennas are more than adequate. You will need a radio that can operate split frequency on separate bands (most modern rigs can do this), or separate radios for 15 and 10 meters.

may be erratic when the satellite is not in full view of the Sun. RS-15 is at it best when used during daylight hours. Activity tends to be light with a mix of CW and SSB.

RS-12 is known as the "DX satellite." Although its orbit is no higher than RS-10, it has the advantage of using a 15-meter uplink. Signals that would normally be out of range of, say, RS-10, can refract through the ionosphere and be received by RS-12. In fact, many hams have been able to work through RS-12 when it is just below their local horizons! This is an impossible feat with RS-10 or RS-15. Thanks to this over-the-horizon characteristic of RS-12, several hams have earned their DXCC awards through this bird!

RS Operating Techniques

Since the satellites are available for a relatively short time, contacts tend to be rather brief. CW operators congregate in the lower half of the transponder passband while SSB operators occupy the upper half.

If you hear someone calling CQ on SSB, note the downlink frequency and quickly tune your transmitter accordingly. As you answer the call, adjust your transmitter until your voice is clear and audible on the downlink. (*If* you're operating full duplex.) You can even do this while he is still calling CQ. I've heard some SSB operators adjusting their uplink frequency and saying, "Test, test, test..." By using this method they're assured of being on-frequency and ready to respond when the other station stops calling.

Answering a CW call is just as easy. As soon as you copy the call sign, tune your transmitter to the proper frequency and start sending a series of dits. Listen on the downlink and adjust your transmitter until you hear the tone of your CW signal roughly matching the tone of the station sending CQ.

If you're using a 2-meter FM rig as your CW uplink transmitter for RS-10 or RS-15, you're probably limited to tuning in 5-kHz steps. This makes it difficult for you to tune onto other stations when they're calling CQ. In this situation it's often best to simply stay in one place and call CQ yourself. The hams who own the more "agile" radios will come to you! Don't attempt to adjust your uplink to compensate for Doppler. That's nearly impossible to do in 5-kHz jumps.

The future of the RS satellites will depend on the level of funding required to keep them operational. Considering the economic hardships facing the Russian Federation at the time of this writing, there is some cause for concern. With any luck, these satellites will continue functioning, providing many enjoyable contacts for years to come.

AMRAD-OSCAR 27: AN FM REPEATER IN ORBIT

OSCAR 27 is basically an FM repeater in outer space. Its output power is only 4 W, but that's plenty when you consider the height of OSCAR 27's antenna!

OSCAR 27 is popular among hams who own dual-band FM transceivers—especially if the 440-MHz section offers receive coverage down to 436 MHz. (Not all dual-banders will receive at this frequency. Check your transceiver manual.) To make contacts through OSCAR 27, you simply transmit on 145.850 MHz and listen on 436.800 MHz (*Mode J*). You don't need beam antennas to work this satellite. Some hams have even managed to make contacts using hand-held rigs and rubber-duck antennas! Of course, the better your antenna, the better your odds of success.

Because only one station at a time can talk through the satellite, contacts tend to be *very* short! OSCAR 27 is only in view for about 20 minutes at best, so lots of hams jump on the air and try to make contacts.

If everyone is polite and patient, you'll be able to make a quick contact, tell the other station where you are, and then say "good-bye." Human nature being what it is, what you'll probably hear instead is a distorted squeal from the satellite as too many stations transmit at once. In this situation "survival of the fittest" rules the radio jungle. He who has

The AMRAD-OSCAR 27 satellite about a month prior to launch.

the most power and the biggest antenna will have the clearest signal. If you're lucky, the big-signal station will appoint him or herself as a kind of "net control" and direct the flow of contacts during the pass. If not, it's a radio slugfest!

OSCAR 27 is only available during daylight hours. OSCAR 27 shares space and power with another on-board system known as *EyeSat*. When it comes to power resources and time, EyeSat gets top priority.

FUJI-OSCAR 20

OSCAR 20 is right on the edge of what most hams would call an EasySat. It's easy from an operational standpoint, but somewhat more difficult when it comes to equipment.

Like the RS birds, OSCAR 20 features a linear transponder. It repeats all conversations it hears taking place in its uplink passband. Most stations working through OSCAR 20 use SSB, although some CW signals are heard as well. With its relatively high orbit, OSCAR 20 offers excellent coverage. For example, hams on the East Coast of the United States can easily work Western Europe when OSCAR 20 is above the mid-Atlantic.

You'll need a 2-meter multimode transceiver to send signals to OSCAR 20 and a 440-MHz multimode transceiver (or simply a 440-MHz SSB receiver) to hear the downlink (see Figure 5-6). This is a substantial investment for most amateurs. In addition, beam antennas are best for this bird, along with an azimuth/elevation antenna rotator to provide horizon-to-horizon tracking. (More money!)

As a result, you don't hear many signals on OSCAR 20. Even so, investing in an OSCAR 20 station will pay off down the road when it comes time to work more sophisticated birds.

A Japanese rocket carries Fuji-OSCAR 20 to orbit.

THE PACSATS

If you enjoy packet operating, you'll love the PACSATs! Several satellites comprise the PACSATs: AMSAT-OSCAR 16, Lusat-OSCAR 19, UoSat-OSCAR 22, KITSAT-OSCAR 23, KITSAT-OSCAR 25 and ITAMSAT-OSCAR 26.

Most PACSATs work like temporary mail-boxes in space. You upload a message or some other file to a PACSAT and it is stored for a time (days or weeks) until someone else—possibly on the other side of the world—downloads it. Neat, isn't it?

Which PACSAT is Best?

You can divide the PACSATs into two types: The 1200 and 9600-baud satellites. OSCARs 16, 19 and

Budgeting Your Satellite Station

The following price estimates are based on *new* equipment. You can expect to save 50% or more if you explore the used-equipment market—and I encourage you to do so.

Note that many of the same components can be used for more than one satellite. The list also does not include other items such as coaxial cable.

The ICOM IC-820H transceiver debuted in 1995. It offers FM, SSB and CW capability on 2 meters and 70 cm.

DOVE

- 2-meter FM transceiver ($500)
- TNC ($120)
- Omnidirectional antenna ($30)

Mir

- 2-meter FM transceiver ($500)
- TNC ($120)
- Omnidirectional antenna ($30)

SAREX

- 2-meter FM transceiver ($500)
- TNC ($120)
- Omnidirectional antenna ($30)

RS-10 and RS-15:

- 10-meter receiver or transceiver ($200 to $1000)
- 2-meter multimode transceiver ($1000)
- Omnidirectional antennas ($100)

RS-12

- HF SSB/CW transceiver ($1000 +)
- Wire antennas ($50)

OSCAR 27

- Dual-band FM transceiver (must be able to receive down to 436 MHz) ($700)
- Dual-band omnidirectional antenna ($100)

OSCAR 20

- 2-meter multimode transceiver ($1000)
- 440-MHz multimode receiver or transceiver ($500 to $1000)
- 440-MHz receive preamplifier—*installed at the antenna* ($150)

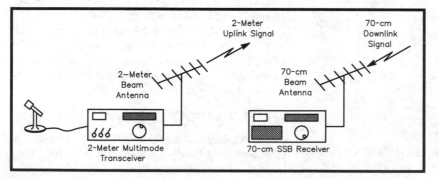

Figure 5-6—A station equipped for OSCAR 20 requires a 2-meter multimode transceiver, a 435-MHz SSB receiver, and beam antennas for the uplink and downlink signals.

The Yaesu FT-736 is the workhorse of the amateur satellite world. Although the design is several years old, it is still very popular. The FT-736 provides multimode operation on several VHF and UHF bands.

- Beam antennas for 2-meters and 440 MHz ($300)
- Azimuth/elevation antenna rotator ($500)

1200-baud PACSATs
- PSK satellite TNC ($300)
- 2-meter FM transceiver ($500)
- 440-MHz multimode receiver or transceiver ($500 to $1000)
- Dual-band omnidirectional antenna ($100)

9600-baud PACSATs
- 9600-baud TNC ($200)

- 2-meter FM transceiver with 9600-baud capability ($500)
- 440-MHz FM transceiver with 9600-baud capability ($500)
- Dual-band omnidirectional antenna ($100)

OSCARs 10 or 13
- 2-meter/440-MHz multimode transceiver ($1500)
- 100-W 440-MHz power amplifier and power supply ($500)
- 2-meter receive preamplifier—*installed at the antenna* ($150)
- Beam antennas for 2 meters and 440 MHz ($300)
- Azimuth/elevation antenna rotator ($500)

The PacComm PSK-1T is a packet TNC designed for operation with PACSATs that use phase-shift keying (PSK) on their downlink signals.

26 are 1200-baud PACSATs. You transmit packets to them on 2-meter FM and receive on 437-MHz SSB. OSCARs 22, 23 and 25 are the 9600-baud PACSATs. You send packets to them on 2-meter FM and receive on 435-MHz FM.

So which PACSATs are best for beginners? There's no easy answer for that question. You can use any 2-meter FM transceiver to send data to a 1200-baud PACSAT, but getting your hands on a 435-MHz SSB receiver (or transceiver) could put a substantial dent in your bank account (see Figure 5-7). In addition, you need a special PSK (*phase-shift keying*) terminal node controller (TNC). These little boxes are not common and could set you back about $250.

So the 9600-baud PACSATs are best for the newbie, right? Not so fast. It's true

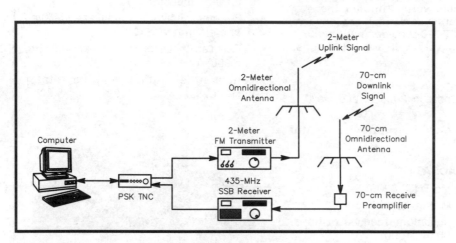

Figure 5-7—This is a diagram of a typical 1200-baud PACSAT station. Notice the special PSK TNC and the 435-MHz SSB receiver. For improved performance, substitute beam antennas.

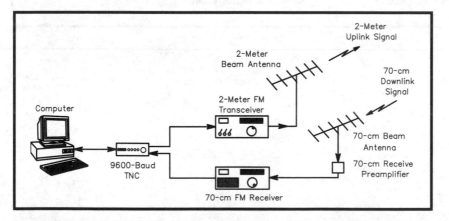

Figure 5-8—A 9600-baud PACSAT station requires less specialized equipment, but the radios must be capable of handling 9600-baud data signals.

that you don't need a special packet TNC. Any of the affordable 9600-baud TNCs will do the job (see Figure 5-8). The catch is that not any FM transceiver is usable for 9600-baud packet. Read the discussion of 9600-baud packet in Chapter 2 and you'll see what I mean. *Both* the 2-meter *and* 440-MHz FM radios must be capable of handling 9600-baud signals. And not all 440-MHz FM rigs can receive down to 435 MHz.

Broadcasting Data

Despite the huge amounts of data that can be captured during a pass, there is considerable competition among ground stations about exactly *which* data the satellite should receive or send! There are typically two or three dozen stations within a satellite's roving footprint, all making their various requests. If you think this sounds like a recipe for chaos, you're right.

The PACSATs produce order out of anarchy by creating two *queues* (waiting lines)—one for uploading and another for downloading. The upload queue can accommodate two stations and the download queue can take as many as 20. Once the satellite admits a ground station into the queue for downloading, the station moves forward in the line until it reaches the front, whereupon the satellite services the request for several seconds.

For example, let's say that OSCAR 16 just accepted me, WB8IMY, into the download queue. I want to grab a particular file from the bird, but I have to wait my turn. OSCAR 16 lets me know where I stand by sending an "announcement" that I see on my monitor. It might look like this:

WB8ISZ AA3YL KD3GLS WB8IMY

WB8ISZ is at the head of the line. The satellite will send him a chunk of data, then move him to the rear.

AA3YL KD3GLS WB8IMY WB8ISZ

Now there are only two stations ahead of me. When I reach the beginning of the line, I'll get my share of "attention" from the satellite.

Unless the file you want is small, you won't get it all in one shot. If the satellite disappears over the horizon before you receive the complete file, there's no need to worry. Your PACSAT software "remembers" which parts of the file you still need from the bird. When it appears again, your software can request that these "holes" be filled.

And while all of this is going on, *you're receiving data that other stations have requested!* That's right. Not only do you get the file you wanted, you also receive a large portion of the data that other hams have requested. You may receive a number of messages and files without transmitting a single watt of RF. All you have to do is listen. That's why they call it "broadcast" protocol.

Station Software

You must run specialized software on your station PC if you're going to enjoy any success with the PACSATs. If your computer uses DOS only, you need a software package known as *PB/PG*. *PB* is the software you'll use most of the time to grab data from the satellite. *PG* is only used when you need to upload.

If you're running Microsoft *Windows* on your PC, you'll want to use *WISP*. *WISP* is a *Windows* version of *PB/PG* that includes such features as satellite tracking, antenna

```
 MSPE [KO-23]                                                    ▼ ▲
 File  Setup  Directory  Fill  Satellite  Send Msg                Help
 12C7A  85   0      0   │ File 12ECB downloaded
 12DE9  29   0      0   │ File 12E9F heard
 12EBA  10  37  85408   │ Saving file 12E87
 12E28   1   0      0   │ Auto: Request fill of file 12E28
 12CE0   2   0      0   │ NO -1 UE1COR
 12E9A   9   0      0   │ Broadcast queue full.
 12E9F   9   0      0   │ File 12CE0 heard
 12D03  16   0      0   │ File 12C7A heard
 12CF4   5  20 106433   │ Auto: Request fill of file 12E28
 12D53  44   0      0   │ File 12CE0 heard
                        │ File 12EBA heard
                        │ Auto: Request fill of file 12E28
 Sat Apr 01 22:42:23 1995 Up: 54/21:7 EDAC: 717 F:78960 L:77648 d:0 s:0 b:2072052
 Sat Apr 01 22:42:38 1995 Up: 54/21:8 EDAC: 717 F:78192 L:77648 d:0 s:0 b:2072061
 Sat Apr 01 22:42:53 1995 Up: 54/21:8 EDAC: 717 F:78192 L:77648 d:0 s:0 b:2072069
 Sat Apr 01 22:43:08 1995 Up: 54/21:8 EDAC: 717 F:78192 L:77648 d:0 s:0 b:2072077
 Sat Apr 01 22:43:23 1995 Up: 54/21:8 EDAC: 717 F:78192 L:77648 d:0 s:0 b:2072088
 Sat Apr 01 22:43:38 1995 Up: 54/21:9 EDAC: 717 F:77488 L:77488 d:0 s:0 b:2072099
 Sat Apr 01 22:43:53 1995 Up: 54/21:9 EDAC: 717 F:77488 L:77488 d:0 s:0 b:2072109
 Sat Apr 01 22:44:08 1995 Up: 54/21:9 EDAC: 717 F:77488 L:77488 d:0 s:0 b:2072121
 Sat Apr 01 22:44:23 1995 Up: 54/21:9 EDAC: 717 F:77488 L:77488 d:0 s:0 b:2072130
 Sat Apr 01 22:44:38 1995 Up: 54/21:10 EDAC: 717 F:77488 L:77488 d:0 s:0 b:207214
 Sat Apr 01 22:44:53 1995 Up: 54/21:10 EDAC: 717 F:77488 L:77488 d:0 s:0 b:207215
 Sat Apr 01 22:45:08 1995 Up: 54/21:10 EDAC: 717 F:78240 L:78240 d:0 s:0 b:207215
 Sat Apr 01 22:45:23 1995 Up: 54/21:10 EDAC: 717 F:78240 L:77488 d:0 s:0 b:207216
 Sat Apr 01 22:45:38 1995 Up: 54/21:11 EDAC: 717 F:77488 L:77488 d:0 s:0 b:207218
 Sat Apr 01 22:45:53 1995 Up: 54/21:11 EDAC: 717 F:77488 L:77488 d:0 s:0 b:207218
 Sat Apr 01 22:46:23 1995 Up: 54/21:11 EDAC: 717 F:77488 L:77488 d:0 s:0 b:207220
 Sat Apr 01 22:46:38 1995 Up: 54/21:12 EDAC: 717 F:77488 L:77488 d:0 s:0 b:207221
 Sat Apr 01 22:46:53 1995 Up: 54/21:12 EDAC: 717 F:78240 L:77488 d:0 s:0 b:207222
 PB: WB1HBU UE3BDR ON6XYᴺD K8TL DC8AM K4UPL ON4KUIᴺD WB2YLR    Open   2a : GU6EFB
 WA8EBM WB2REM NZ3FᴺD DF5DP GØSUL F1TIU G3CDKᴺD LX1BB UE3EGO
 G3ZUMᴺD UE3FRH
 │ DIR 44 holes │ AUTO 12E28 │ 83% │ T:540631 │ D:40934 │ F:375225 │
```

WISP software in action as it communicates with the KITSAT-OSCAR 23 packet satellite. In the lower left portion of the screen you can see the list of stations waiting in the queue.

rotator control and more. Both software packages are available from AMSAT at the address mentioned elsewhere in this chapter.

It's only fair to warn you that both software systems are somewhat complex from a user standpoint. *PB/PG* has a reputation for being "user hostile." The on-screen messages are cryptic until you finally figure them out—usually on your own. (No, the documentation won't be of much help.) Installation is tricky because you must create various directories and subdirectories for each satellite. *WISP* is friendlier, but *Windows* experience is a big asset. Getting the software to talk to external devices such as the TNC and antenna rotators may cause major headaches for some hams.

I'm not trying to scare you away from the PACSATs—far from it! Just be prepared to tackle a steep learning curve, depending on your computer experience. Don't let frustration get the better of you. In time, and maybe with a little outside assistance, you *will* get on the PACSATs . . . and you'll be glad you did.

OSCAR 13 AND OSCAR 10

Throughout this chapter we've been talking about satellites that travel in low-Earth orbits. The advantages of these satellites are obvious: you can access them with relatively low power and very meager antennas. On the other hand, they are available only for brief periods of time and the opportunities for DX (long-range communication) are almost nonexistent.

Fortunately, there are two satellites that travel in high, elliptical orbits and they're both DX powerhouses: AMSAT-OSCAR 13 and AMSAT-OSCAR 10.

As you'll see in Figure 5-9, both satellites have orbits that act like slingshots, shooting them out to altitudes of more than 30,000 km. At the high points of their orbits, they seem to be nearly motionless from our perspective here on Earth. While a certain amount of antenna aiming is required, very little additional movement is necessary once the antennas are in their proper positions. From their high vantage points, OSCARs 10 and 13 "see" a great deal of the Earth. This opens a window to DX contacts on a regular basis!

The downside to having such a high-altitude orbit is that more transmitted power is needed to access the satellites and a weaker signal is received. You'll need high-gain, directional antennas to operate OSCARs 10 or 13. The more gain you have at the antenna, the less power will be required at the transmitter. You'll need to be able to

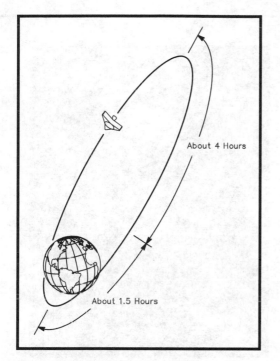

About 4 Hours

About 1.5 Hours

Figure 5-9—OSCARs 10 and 13 travel in high, elliptical orbits that shoot the spacecraft far out into space.

rotate the 2-meter and 70-cm antennas vertically *and* horizontally (elevation and azimuth).

Mode B (70-cm uplink/2-meter downlink) is the most popular mode on OSCARs 10 and 13. (See the frequencies listed in Table 5-1.) A 2-meter SSB/CW receiver is required for the downlink and a similar 70-cm transmitter is necessary for the uplink. Considering the weak signals, a 2-meter mast-mounted preamplifier is also a worthy addition to your station.

Like the RS satellites, OSCARs 10 and 13 employ linear transponders to "repeat" many signals at once. SSB and CW are the modes of choice, although conversations tend to be longer and more relaxed. With these satellites you don't have to worry too much about losing the signal in the middle of your conversation!

And Now the Bad News

OSCAR 13 is a doomed satellite and OSCAR 10 is crippled. How's that for a blunt summary?

OSCAR 13 has a slight problem with its orbit. Every time it swings in close to the Earth, gravity and the atmosphere tug at the bird, reducing its inertia. As a result, it drops even lower on the next pass . . . and the next . . . and the next. By the summer of 1996, OSCAR 13 will be churning through enough atmosphere to cause significant heating. Some onboard systems will probably begin to fail. By December 1996, the Earth will finally win the struggle and OSCAR 13 will become little more than an ash cloud drifting in the upper atmosphere. If you're reading this book in January 1997 or later . . . uh . . . never mind.

OSCAR 10 is in a stable orbit, but it has had a rough life. The poor satellite has been hammered by high levels of radiation and they've taken their toll. It's out of control in the sense that it doesn't respond to orders from the command stations. (They gave up long ago.) It simply cruises through space, repeating whatever signals it happens to pick up. On some days, OSCAR 10 is terrific. It will sound every bit as good as OSCAR 13. On other days, you can barely hear it. Eventually the bird will go silent for good, but no one knows when that might happen.

THE FUTURE—PHASE 3D

The hopes of many OSCAR 10 and 13 users are riding on *Phase 3D*. The satellite is scheduled for launch in late 1996, or early 1997. If Phase 3D reaches orbit safely (everyone is crossing their fingers!), the result will make OSCARs 10 and 13 look like peanut whistles.

Phase 3D is, without a doubt, the largest, most expensive Amateur Radio satellite ever created. It's an international effort with components supplied by

Satellite Antennas

As with any Amateur Radio station, your antennas determine how well you'll communicate. Connecting an expensive radio to a lousy antenna is a sure-fire formula for frustration!

The *eggbeater* is an omnidirectional antenna that creates a circular radiation pattern overhead. This is ideal for satellite applications.

A close-up view of an azimuth/elevation antenna rotator. This device will move your antennas from side to side (azimuth) as well as up and down (elevation). The small boxes immediately below the rotator are receive preamplifiers.

Directional (beam) antennas such as Yagis or quads are best for any satellite station. They concentrate your transmitted and received energy, allowing you to more easily bridge that multihundred or multi*thousand* mile gap between you and the satellite.

The main problem with beam antennas, however, is cost and space. Beam antennas are not cheap, and you'll need expensive azimuth/elevation rotators to use them properly. Az/el rotators spin the antennas side to side as well as up and down. A new unit typically costs $500. Beam antennas and rotators also require space on your roof or wherever. Not every ham has the room to install such a system.

So, some satellite-active hams must compromise and use omnidirectional antennas. These antennas are more affordable, fit into relatively tight spaces and do not require rotators. On the other hand, they have no signal-concentrating ability.

The Importance of Coax

Regardless of which antenna system

Classic cross-polarized Yagi antennas like these are the ultimate performers at any satellite station. As you can see from this photograph, they don't necessarily need to be installed on rooftops, either! The larger beam in the foreground is used for the 2-meter downlink for OSCAR 13. The small beam in the background is the 70-cm uplink antenna.

you select, be sure to connect them to your radios with the best coaxial cable you can afford. Don't cut corners here! Take a look at the coax table in Chapter 2 and follow those guidelines. As a rule of thumb, you can use Belden 9913 coax (or equivalent) for all your satellite antennas. You may also find *LNR* coax that has similar low-loss characteristics. Either 9913 or one of the LNR varieties are good choices for the satellite stations we've discussed in this chapter.

Polarization?

In Chapter 2 I said that antennas for FM communication must be *vertically* polarized. In Chapter 3 I stated that antennas for SSB or CW on the VHF/UHF bands must be *horizontally* polarized. What type of polarization is best for satellite work?

For most satellites, antenna polarization is not quite so critical. If you're using an omnidirectional antenna, vertical polarization is fine. You may encounter a few omni antennas that offer *circular* polarization (such as the so-called "eggbeater" designs). These are even better for satellites, but they cost substantially more.

If you browse through the ham-equipment catalogs, you'll find *cross-polarized* beam antennas for satellite stations. These are bascially Yagi antennas with both horizontal *and* vertical elements. A remote-control relay allows you to select one set of elements or the other.

Do you need cross-polarized beams? For the high-altitude birds such as OSCARs 13 and 10, a cross-polarized beam can give you a decided signal advantage. For the other satellites, their value is questionable. A set of vertically polarized beams often works just as well.

Antenna System Recommendations

RS-10, RS-15: Omnidirectional antenna for the 2-meter uplink. Wire dipole antenna for the 10-meter downlink

RS-12: A multiband wire dipole antenna (tuned for 15 and 10 meters, among other bands)

Mir, SAREX: 2-meter omnidirectional antennas

OSCAR 27: 2-meter/440-MHz dual-band omnidirectional antenna

OSCAR 20: 8-element, 2-meter beam and a 10-element, 440-MHz beam with an azimuth/elevation rotator

1200-baud PACSATs: 2-meter/440-MHz dual-band omnidirectional antenna

9600-baud PACSATs: 8-element, 2-meter beam and a 10-element, 440-MHz beam with an azimuth/elevation rotator

OSCARs 10 and 13: 22-element, 2-meter beam and a 38-element, 440-MHz beam with an azimuth/elevation rotator. Both beams cross-polarized.

Dick Jansson, WD4FAB (left), and Steve Ford, WB8IMY, inspect the partially assembled Phase 3D satellite as it's being assembled in a "clean room" on the grounds of the Orlando, Florida, International Airport.

amateur satellite organizations and individuals throughout the world. Take a look at Figure 5-10 and you'll see how the size of Phase 3D compares to other ham satellites. Those sections that resemble wings are actually solar panels. They unfold to provide power to the satellite after it reaches orbit.

Phase 3D channels this power to run an incredible array of transmitters and receivers on frequencies from 21 MHz to 24 GHz. The RF output of its 2-meter transmitter alone will be about 200 W. Compare that to the 50-W output of OSCAR 13 on 2 meters. This isn't the whole story, though. OSCAR 13's 2-meter antenna offers an effective radiated power (ERP) of 180 W. The superior 2-meter antennas aboard Phase 3D are capable of yielding an ERP of up to *2500 W!*

What does this mean to you? It means that you won't need the large multielement beam antennas you're accustomed to seeing on most OSCAR 10 and 13 stations. Depending on how sensitive your receiver is, you may not even need a mast-mounted receive preamp.

Phase 3D will pack a substantial punch on all of its transmitters. The same is true for the sensitivity of its receivers. Large, expensive satellite stations will be a thing of the past. This will be especially true if you take advantage of Phase 3D's microwave capability.

Oh No! Not Microwaves!

Yes, microwaves. Phase 3D will really shine on the bands above 1 GHz. Let's face it, the future of amateur satellite communication is microwave. On the microwave bands you can get superb gain from tiny antennas. Noise levels are extremely low, too. Phase 3D's microwave ability will make it possible to install your satellite station in a cramped attic with room to spare—or on the balcony of your apartment building. Your neighbors won't even know your station exists.

But microwave equipment is difficult to use or build, right? Wrong. The days of microwave "plumbing" projects are behind us. Now we have *transverters* that will take 2 or 10-meter signals and turn them into 1.2, 2.4, 10 or 24-GHz satellite uplinks. You can buy these transverters or build them. (Yes, they're available as kits.)

Good-bye Tracking Software?

OSCAR 13 travels in a high, elliptical orbit. So will Phase 3D. Because of the nature of OSCAR 13's orbit, however, it doesn't appear in the same part of the sky on a regular basis. You need tracking software to tell you when OSCAR 13 will be within range of your station, and where you should point your antennas.

There is an extra dimension to the tracking problem with OSCAR 13, though. Every few months, OSCAR 13's attitude in space must change so that all of its solar cells receive the proper amount of sunlight. This can cause a great deal of

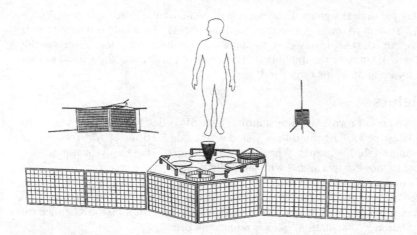

Figure 5-10—Compared to other Amateur Radio satellites, Phase 3D is a monster!

frustration if you haven't taken the attitude factor into account. Your computer might tell you that the satellite is directly overhead, but when you switch on your radio, you hear only weak signals or nothing at all. Why? Because OSCAR 13's antennas are pointing away from your station!

Phase 3D will do away with most of these headaches. A complex system of magnets, sensors and magnetically suspended *reaction wheels* will keep the spacecraft and its antennas pointed at the Earth at all times. With its specially designed solar panels, Phase 3D doesn't need to change attitude to get sufficient sunlight.

And Phase 3D's orbit will place the satellite at the same position above your horizon every 48 hours. If the satellite is, say, 45° above your southwest horizon at noon Sunday, it will be there again at noon Tuesday. This feature of Phase 3D is a blessing to hams with shallow pockets. If you don't mind not being able to use the bird for the full time it's available, you could aim your antennas at a particular point in the sky and leave them there. No expensive azimuth/elevation rotators necessary. Of course, I wouldn't throw away your tracking software yet, but its importance for working Phase 3D will be greatly diminished.

If this wasn't enough, you may be surprised to learn that Phase 3D may be able to *provide its own orbital elements*. The satellite is expected to carry a *global positioning system* (GPS) receiver that will continually monitor Phase 3D's position in space with a high degree of accuracy. This information can then be translated into the usual Keplerian format and sent down to you as telemetry. Whenever you want to update your software, you can get the data directly from the horse's mouth, so to speak.

Packet, Too

While SSB and CW will be the primary modes of communication through Phase 3D (sorry, the power budget won't permit the use of FM), the satellite will also function as a relay for digital signals. Packeteers will be able to take advantage of Phase 3D's *RUDAK* system to transfer data over vast distances. The nominal data rate will be 9600 bit/s, but RUDAK is designed to operate at much higher rates. Some see the day when Phase 3D may be the linchpin in a network of packet satellite gateways that will communicate at 56 kbit/s or beyond.

Bulletins on 10 Meters

Even if you don't own a satellite station, Phase 3D is designed to keep you up to date on the latest news. The satellite is slated to include a 10-meter AM transmitter with approximately 200 W output. When Phase 3D is close to the Earth (near perigee), you'll be able to monitor this signal with an ordinary short-wave receiver. You'll hear all sorts of voice announcements about the status of the satellite, news about other satellites, or Amateur Radio news in general. Imagine relaxing in your backyard as you listen to a portable radio. When your neighbor asks what station you've tuned in, you reply, "Station? What station? This is a ham satellite!"

The Camera Never Lies

When the world crashes into my living room,
Television man made me what I am.
People like to put the television down,
But we are just good friends.
I'm a television man.

—The Talking Heads, *Television Man*

"Hams . . . and the people who share their secret desires! It's all coming up today on the AA6XYZ show!"

(Switch to wide-angle pan. Show audience of neighborhood kids, dog, goldfish and petulant spouse.)

(Dissolve to call sign logo.)

(Fade to black.)

Okay, not all hams make "productions" out of their amateur television transmissions, but it's tempting, isn't it? After all, this is *your* television station. You call the shots.

I bet the idea of operating a television station from your home probably makes you want to hide your checkbook and credit cards in a safe place. That's understandable. Amateur television (ATV) must surely be the most expensive aspect of our multifaceted hobby, right?

Uh . . . no.

Depending on how fancy you want to get, you can put your own ATV station on the air for hundreds of dollars *less* than the cost of a new 2-meter multimode transceiver.

And ATV can be operated by *anyone* with a ham license. Novices, for example,

Table 6-1

Popular 70-cm ATV Frequencies (MHz)

Video	Audio	Use
421.25	425.75	repeater outputs
426.25	430.75	repeater inputs or outputs
434.00	438.50	simplex and repeater inputs
439.25	443.75	simplex, repeater inputs and outputs

have ATV privileges from 1270 to 1295 MHz. Hams licensed at the Technician class or higher enjoy maximum flexibility. They have full access to all amateur frequencies above 50 MHz, and most ATV is in the 420-440 and 902-928 MHz bands. The 70-cm band (420 to 440 MHz) is the most popular ATV band by far (see Table 6-1).

THE MANY FLAVORS OF ATV

Television operators are a tight-knit group because they are relatively few in number (compared to other segments of the hobby). Aside from the occasional band openings that send their signals bouncing over hundreds of miles, most ATV conversations are local.

It's not uncommon for ATV stations to take to the airwaves in the evening hours to share the news of the day. They may simply chat in *roundtable* format, each operator taking his or her turn before the camera.

"Look at this tomato," KE5ABC says as he holds the rotund vegetable before the lens. "I plucked it off the vine just after dinner. Isn't it a beauty?"

If you watch such a roundtable, you may also see videos of a local club's Field Day operation, weather data or just about anything you can think of (as long as it's not commercial or obscene,

Richard Logan, WB3EPX, enjoys operating ATV from his home station.

Radio-controlled airplane flying was never like this! With a tiny camera behind the nose gear and an ATV transmitter in the fuselage, you can view the world as your aircraft sees it! (*photo by Ron Berkman, KA9CAP*)

Students at Southeastern Community College in Whiteville, North Carolina, prepare to launch a sophisticated ATV-equipped rocket!

of course!). Other ATVers are more technically oriented. They meet on the air to compare software, antennas, cameras and so on.

Keep in mind that "tight-knit" does *not* necessarily mean "exclusive." ATV operators are delighted to have new stations join the group. In fact, many TV enthusiasts spend a good deal of their time encouraging other hams to test the waters.

Avid ATVers enjoy an extraordinary range of operating exploits. You can mount a simple camera, transmitter and associated electronics in a compact package and launch it as the payload beneath a helium balloon. Experimenters have amassed thrilling videotape recorded from the received signals of ATV-equipped balloons at the edge of space, more than 100,000 feet up! The view from an altitude of 20+ miles is breathtaking.

Closer to the ground, hams combine hobbies by controlling their radio-controlled aircraft, boats and cars with a "pilot's-eye view" monitored from a camera mounted in their miniature craft. Amateur rocketeers mount ATV equipment inside model rocket nosecones for a rapid ride up to 1000 feet or more, watching the world recede beneath their "eye in the sky."

Local public service agencies and government emergency preparedness offices are enthusiastic about having experienced hams provide television coverage of community events, drills and disasters. You can take your portable ATV station aboard a police or National Guard helicopter to survey the extent of forest fires, floods and storm damage. Parades, walkathons, bicycle races and other outdoor activities can be supported by skillful placement of mobile ATV units sending video images to command posts. Trained weather observers can aid National Weather Service officials by letting them see developing storms or tornadoes firsthand.

Get a Glimpse of ATV

Before you dive head first into ATV, it's a good idea to check out the activity in your area. If you live in or near a large city, you have a decent chance of finding some ATV stations on the air. In rural locations, however, ATV operators may be hard to find.

Many ATV operators rely on dedicated ATV repeaters to extend their range. ATV repeaters are similar to voice repeaters (see Chapter 2), except that they relay television signals. Like voice repeaters, ATV repeaters are usually placed on hilltops, building roofs or other high-altitude locations. They also use a considerable amount of RF output power.

Because of their strong signals, ATV repeaters are relatively easy to monitor. Get

Table 6-2

Cable TV Channels by Radio Frequency

Note: These are *cable* channels, not UHF TV channels.

Channel	MHz
57	421.25
58	427.25
59	433.25
60	439.25

your hands on a recent copy of *The ARRL Repeater Directory* and you'll find a list of ATV repeater frequencies. Another good source of information is your local packet-radio network. Simply send a local bulletin asking for active ATV frequencies in your area.

If you think you have ATV activity nearby, the next step is to watch it! Many portable LCD televisions can tune ATV frequencies directly. The same is often true of cable-ready TVs and VCRs. To watch ATV with your own receiver, all you have to do is tune to the appropriate "channel" (see Table 6-2).

If you don't mind spending about $100 to test the waters, you can purchase an ATV *downconverter* from one of the vendors listed in the Info Guide. A downconverter takes 70-cm ATV signals and converts them "down" to channel 3 or 4. All you need then is a conventional TV to watch the fun.

Unless you live in the shadow of a powerful ATV station or repeater, you'll need a beam antenna to receive a watchable picture. Try a small, portable beam connected to your receiver through a short piece of low-loss coax. Point the beam out a nearby window and do your aiming by hand (this is just an experiment, after all).

The cheapest and easiest way to get your first glimpse of ATV is through another ham's station. Start asking around at the next club meeting, or on your local repeaters. I'll wager that you'll find at least one ham who is more than happy to invite you over to see his or her station.

PLANNING YOUR ATV STATION

Unless you choose to set up the ultimate station, ATV is very affordable (see Figure 6-1). If you happen to have a typical home video camera or camcorder and a basic TV set (black and white is fine), you can purchase an ATV transceiver for about $400. Add perhaps $75 for a reasonable antenna and a few dollars for cables and miscellaneous odds and ends.

Antennas, Amplifiers and Preamplifiers

Don't attempt ATV with anything other than a beam antenna. An ATV signal is several *megahertz* wide. This means that your RF output is being spread over a broad chunk of spectrum. The more you spread RF power, the less punch it has on the receiving end. (That's why commercial TV stations often run several *megawatts* of effective radiated power!) If you expect reasonable range, you must focus your signal as much as possible with a beam antenna.

The good news is that a high-gain multielement Yagi for 434 MHz is only two or

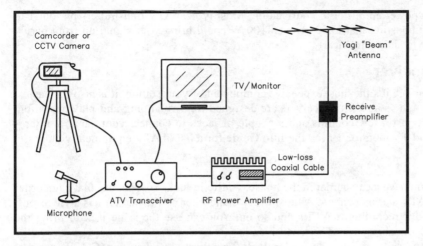

Figure 6-1—A diagram of a typical ATV station.

three feet long. Antennas for 902 and 1296 MHz are even smaller. Yes, you'll need a rotator to aim your antenna, but an inexpensive TV antenna rotator ($60 or less) is more than adequate. If you live in an apartment or condominium, a small beam and a rotator will fit nicely in almost any attic.

When choosing your ATV antenna, don't forget about *polarization*. Most antennas used in ATV applications are either horizontally or vertically polarized. Yagi antennas, for example, can be horizontal or vertical depending on how you mount them. Using the proper polarization is important. If you're vertically polarized and the station you're talking to is horizontally polarized, you'll both experience a noticeable loss in received signal strength.

So which antenna polarity should you use? It all depends on the ATV activity in your area. If you intend to operate through an ATV repeater, vertical polarization is best since most repeaters use vertically polarized antennas. For direct contacts—or ATV DXing—horizontal polarization is the standard. Some antenna manufacturers allow you to enjoy the best of both worlds with dual-polarization antennas. These are basically two Yagi antennas mounted on the same boom. One is in the horizontal position and the other is vertical. An antenna switch (often remotely controlled) is used to select one antenna or the other.

If you've read the previous chapters, you can probably guess what's coming next. Yes, it's the low-loss-coax mantra! I've said it several times already, but it must be repeated again: Use the lowest-loss coax that you can afford. When it comes to ATV, you're dealing with UHF signals. Loss can be horrendous at UHF if the SWR is even somewhat elevated. Check the coax table in Chapter 2 and buy accordingly.

In marginal areas, a receiver preamp mounted at the antenna can be a big help (it's probably a good idea no matter where you live). If you have a powerful TV broadcast transmitter nearby, you might have to add some high- or low-pass filters to keep the commercial station from overwhelming your receiver.

If you aren't within a dozen miles or so of a local ATV repeater, think about adding a power amplifier to your station. Most typical ATV transmitters put out 1 to 5 W, and boosting that to 50 or even 100 W could bring your signal up from barely copyable to sharp and clear!

Your ATV Transceiver

Although it's the most expensive component of your station, it is also the easiest to set up and use. ATV transceivers are designed for virtual plug-and-play operation. There are just a few controls and a couple of jacks to connect your camera, microphone and TV monitor. Check the Info Guide for a list of ATV equipment vendors.

The Camera

If you have a camcorder in the house, you're in luck. You can use that camcorder as your ATV station camera! Almost every camcorder on the market has a video-output jack for recording to VCRs and so on. You can use the same jack to bring the camera's signal to your ATV transceiver.

If you can't afford a camcorder, look for a used *closed circuit* (CCTV) camera. Most of these cameras are black and white and their resolution (image detail) isn't the greatest, but they'll be sufficient to get you started. If you search the ham catalogs and the advertising pages of *QST* magazine, you'll often find tiny cameras retailing for $150 or less.

Used CCTV cameras are often found at hamfest flea markets, sometimes for as little as $30. If you buy a used CCTV camera, be prepared to do a little work. You may have to build a special power supply, or you may have to troubleshoot a few problems. Even so, if you purchase a camera that's in decent shape, you'll be on ATV at a bargain price!

YOU'RE ON THE AIR!

How do you call "CQ" on ATV? If you want to be informal, just sit down in front of the camera and make yourself comfortable. Activate your transmitter and say, for example, "This is KD0XYZ in Hannibal, Missouri. Anyone around?" Switch back to the receive mode and wait for a response. If no one replies, choose another frequency (or repeater) and try again.

Another approach is to broadcast an image of your call sign for several seconds while announcing your invitation. Your call sign should be large enough to fill the entire screen. Bold, black letters on a white background will make it easier to read at greater distances (black lettering on white cardboard will do nicely). Computers or character generators can also be used to create

The W5KPZ ATV repeater in Tyler, Texas. Repeaters like these provide wide coverage for even small, low-power ATV stations.

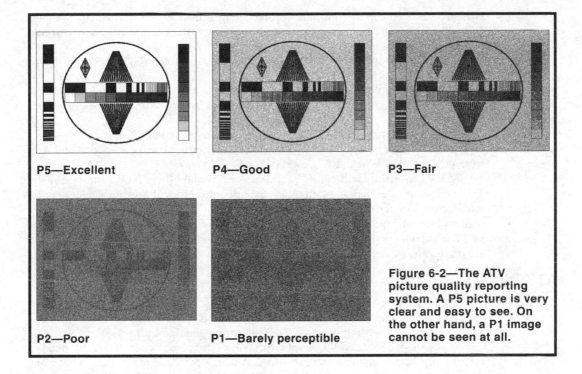

P5—Excellent   **P4—Good**   **P3—Fair**

P2—Poor   **P1—Barely perceptible**

Figure 6-2—The ATV picture quality reporting system. A P5 picture is very clear and easy to see. On the other hand, a P1 image cannot be seen at all.

attractive ID screens. Let your imagination be your guide!

You can establish contact with another station on a single frequency (simplex), or use an ATV repeater with separate input and output frequencies (duplex). The same operating procedures apply in either case. Some ATV repeaters transmit *beacons* (usually composed of its call sign and some graphics) to help you find them. Other ATV repeaters transmit their identifications at regular intervals (on the hour and half-hour, for example). When you transmit, the beacon will disappear and your video signal will be relayed through the system. When you stop transmitting, the beacon will return.

When you finally establish contact, the first order of business is to exchange signal reports so that each of you will know how well your signals are being received. In ATV we use the "P" system to describe the amount of noise (or "snow") in the picture. A P5 picture, for example, is excellent with no visible noise. A P1 picture is very weak with a great deal of noise (see Figure 6-2).

After that, anything goes! You can talk about whatever comes to mind. Unlike other Amateur Radio modes, the person at the other end will be able to see your gestures and facial expressions as you speak. This gives ATV a unique personal dimension that is difficult to achieve on SSB, CW, FM or packet.

The 2-Meter Connection

In many areas of the country, it's a popular practice to use a 2-meter FM simplex frequency—or a repeater—for local ATV coordination. This allows ATVers to make critical adjustments *while transmitting*. Since mutual interference between the 2-meter and 420-MHz equipment is minimal, ATVers can get instant comments on their signal quality and make additional adjustments as necessary.

"AA8QVC from N8SVN. Your signal is getting worse. You turned your beam too far."

"There. The antenna should be pointing directly at the repeater. How's that?"

"Much better, Tom. I can see you and the chair you're sitting in, but everything else is pretty dark."

"That's a lighting problem. Let me switch on my desk lamp."

"Nice! Now I can see everything. Say, that's an impressive movie poster you have on your wall. Where did you get it?"

Many enthusiasts will call CQ on 2-meter FM and 70-cm ATV simultaneously. This technique is effective in attracting the attention of local operators who may be monitoring 2 meters while their ATV equipment is inactive. If ATV coordination is taking place on 2 meters in your area, try to find the frequency. (144.34 MHz is popular.) By listening to the chatter you'll get a pretty good idea of how many ATV stations are on the air, where they are located and when they are active.

ATV DXing

There is more to ATV than chatting with your local friends. You'll also have opportunities to make contacts with ATVers hundreds of miles away. It doesn't happen every day, but when it does, ATV DXing offers genuine excitement!

If you plan to work ATV DX via direct, single-frequency contacts, you'll need to invest in the best equipment you can afford. Antenna-mounted preamplifiers are mandatory along with RF power amplifiers and low-loss coaxial cable or Hardline. Some ATV DXers use a single, high-gain Yagi antenna (horizontally polarized). Others use several Yagis working together in a carefully designed assembly fed by the primary coaxial line. This is often called an *array*. Depending on its construction, an array is capable of very high performance in DX applications.

You can also work DX through your local ATV repeater. When the band is open, many ATVers will attempt to reach repeaters in distant locations. Don't be surprised if you call CQ on your repeater one evening and receive a reply from an ATV DXer!

The best times for ATV DX are around sunrise and right after sunset. If your family TV is connected to an outside antenna, tune through the commercial UHF channels. If you begin to see distant stations, especially on channels 14 through 30, there may be a band opening in progress!

When it comes to video DX, John, KD0LO, is an expert! He has enjoyed ATV contacts with stations more than 400 miles from his home in St Louis, Missouri. (*photo by Bill Brown, WB8ELK*)

Working the World on a Wire 7

You see, wire telegraph is a kind of a very, very long cat. You pull his tail in New York and his head is meowing in Los Angeles. Do you understand this? And radio operates exactly the same way: you send signals here, they receive them there. The only difference is that there is no cat.

—Albert Einstein, when asked to describe radio

Throughout most of this book I've been babbling *ad nauseum* about the exciting world above 50 MHz. That's because this primer is designed at heart for the huge number of hams who enter the hobby through the "codeless" Technician license. For these amateurs, the world above 50 MHz is the only world they know.

But we can't forget the many Technicians who eventually upgrade to Technician Plus, General, Advanced or even Extra. (Or the hams who *begin* their amateur careers with these advanced licenses!) When you step above the Technician class, you're walking through a portal that opens on another world. We're talking about that strange portion of the radio spectrum where the fates of signals are governed by the Sun and a high-altitude layer of our atmosphere known as the *ionosphere*. A signal transmitted on these frequencies can go almost anywhere on the Earth, depending on the fickle nature of propagation. We refer to this fascinating electromagnetic dominion as *High-Frequency*, or simply *HF*.

NINE FLAVORS OF ICE . . . ER, *SPECTRUM*

There are nine groups of frequencies—or *bands*—in the HF/MF region where hams are allowed to operate. Each has its own "flavor." Let's take a brief look . . .

160 Meters: 1.8 to 2 MHz

This is the "basement" of Amateur Radio, the lowest band that we can use. It's almost *un*usable during daylight hours, but nighttime signals on 160 can travel hundreds or even thousands of miles. Noise levels are high, especially during the summer months when thunderstorms march across the landscape. This makes 160 meters primarily a "winter" band. You'll find a mix of CW and SSB activity on this band, although most of the long-distance work (DX) takes place on CW. Because of the extremely long antennas required for 160 operating, it isn't a band for everyone. As a result, crowding is at a minimum.

80 Meters: 3.5 to 4 MHz

Like 160 meters, 80 meters is considered a nighttime band. Even so, 80 is also good for daytime communication out to a few hundred miles.

Depending on atmospheric conditions, 80 meters can offer worldwide communication at night. Even under mediocre conditions, contacts between the East Coast of the US, for example, and Europe are common. Like 160, this band also suffers from high noise levels, so it is definitely at its best during the winter.

The voice portion of 80 meters is inhabited primarily by SSB (single sideband) voice operators, although you'll hear a couple of AM signals here and there. When the band is in decent shape, the voice portion can become extremely crowded. This is particularly true between 3.85 and 4 MHz.

You may also find wall-to-wall signals in the CW segment. Chasing international contacts on 80-meter CW is a favorite pastime. The digital modes (which we'll discuss later in this chapter) can be found on 80 meters, but the activity level is usually light.

40 Meters: 7 to 7.3 MHz

Forty meters is a transition band. It shares characteristics with the lower and higher HF bands.

During the day, 40 is excellent for communication over distances of about 500 miles or so. Phone operators enjoy meeting on 40 for late-morning chats. *Nets*—groups of hams who congregate on the air for a specific purpose—also exploit the advantages of 40 meters in the daylight hours.

At night, 40 meters opens to the world. CW operators enjoy global range in the lower portion of the band. SSB enthusiasts would relish the same contacts were it not for severe interference from shortwave broadcast stations. These high-powered RF blasters decimate the 40-meter voice segment throughout much of the world at night.

The digital modes are popular on 40 between 7.060 and 7.100 MHz, but they face stiff competition from hams in other countries who have *voice* privileges on these frequencies. This interference is especially bad at night.

30 Meters: 10.1 to 10.150 MHz

This is strictly a CW and digital band. No voice operating is allowed. In addition, you can't use more than 200 W output.

Thirty meters is good for DX work during the daylight hours, and up to several hours after dark. The CW operators gather in the lower portion of the band to chase international contacts. Thirty meters is also a terrific band for low-power (QRP) CW activity. The digital folks occupy the upper portion.

20 Meters: 14 to 14.350 MHz

Twenty meters is the queen of the DX bands. It's open to just about every corner of the world at various times of the day. During the peaks of the 11-year solar cycle, 20 meters will remain open throughout the night. (The next peak will take place from about 1999 through 2003.) Otherwise, it tends to shut down after local sunset.

All modes are hot and heavy on 20 meters! SSB reigns supreme in the upper half of the band. You may have a tough time finding a clear frequency between 14.225 and 14.350 MHz, especially on weekends. CW occupies much of the bottom portion. DX chasers often hunt between 14 and 14.050 MHz. Low-power (QRP) CW operators hang out around 14.060 MHz. The digital modes can be heard anywhere from 14.065 to 14.120 MHz.

17 Meters: 18.068 to 18.168 MHz

This daytime band also offers worldwide communication, though it is never crowded. SSB seems to be the dominant mode of communication on 17 meters, although you'll find a few CW stations and the occasional digital operator.

Seventeen meters is best during the peak years of the solar cycle. Even so, it can often provide global DX even in the worst years.

15 Meters: 21 to 21.450 MHz

This is a hot DX band when the solar cycle reaches its peak. In cycle doldrums, 15 meters can provide some mediocre DX, but it's usually limited to sporadic openings over 500 to 1,000 miles.

Like 17 meters, 15 is a daytime band. In the best years, it opens in the late morning and closes a few hours after dark. You'll find a mix of SSB and CW on this band, although the SSB operators usually predominate. The digital segment can also be quite active.

12 Meters: 24.890 to 24.990 MHz

When the solar cycle is up, so is this band—at least in the daylight hours. When the cycle is down, 12 meters is a wasteland. Even in the best years, contacts are few and far between. This has more to do with a lack of interest than anything else. You'll hear the odd SSB and CW conversation, but most operators prefer to migrate to the next-highest band—10 meters.

10 Meters: 28 to 29.700 MHz

During the best years of the solar cycle (coming up from 1999 through 2002), 10 meters is one of the hottest DX bands around. At solar-cycle peaks, the ionosphere absorbs relatively little of your signal at this frequency. It simply bends it back to Earth thousands of miles away. As a result, even low-power stations can use 10 meters to work the world with ease.

When sunspots are scarce, so are contacts on 10 meters. In fact, many hams consider 10 meters to be worthless during the *solar minimum*. That's an exaggeration.

While it's true that you won't make too many DX contacts during the low points in the cycle, the band frequently opens for conversations over hundreds of miles.

When it's open, 10 meters is usually a daytime band. It opens in the late morning and shuts down at dark. SSB activity is most heavily concentrated in the segment from 28.300 to 28.500 MHz. CW is relatively rare and so are the digital modes. At the top end of the band you'll find a segment from 29.500 to 29.700 that's dedicated to FM operating.

SETTING UP YOUR HF STATION

Grab a piece of paper and prepare for a pop quiz. (Or you can write in this book if you dare deface such a valuable piece of literature!)

1. I want my first HF station to operate . . . (check one)
   - ☐ On one band
   - ☐ On several bands
2. I want to operate . . . (check one)
   - ☐ SSB
   - ☐ CW
   - ☐ SSB and CW
   - ☐ Digital
   - ☐ All of the above
3. I don't want to spend more than: _____

Congratulations! You've just established three major criteria for building your first HF station. Mark this page because you'll refer to it often as we get down to brass tacks. Right now, in fact . . .

The Radio

This is the category where sticker shock may hit you the hardest—depending on your answers to questions 1 and 2. New multiband HF transceivers are not cheap. You'll pay at least $1100 for a 100-W transceiver that can do SSB, CW and the digital modes. If you insist on a radio that has every conceivable goodie you can imagine, you could fork over as much as $8000! (But only rich and/or insane hams do this.)

The ICOM IC-707 transceiver is a back-to-basics design.

The Kenwood TS-50S is a compact HF transceiver for home or mobile use.

The Ten Tec *Scout 555*, one of the few under-$1000 rigs, offers single-band CW or SSB at 50-W output. You change bands by swapping modules.

Yaesu's FT-900AT is a full-featured radio for home or mobile work. The detachable front panel makes it easy to hide the main body of the rig in a safe location.

If you have a grand or more lying around, go for the gusto. Today's amateur transceivers are excellent bargains when you consider the level of technology they represent. The radio you buy today will probably still be plugging away decades from now. To make the best purchase decision, read the Product Reviews that appear each month in *QST* magazine. We've also compiled past reviews in our *Radio Buyer's Sourcebooks*.

But if a four-figure price tag jolts you into cardiac arrest, look into *used* equipment. Many amateur dealers who advertise in *QST* stock used gear and even offer warranties of 30 to 90 days. You'll pay a little more, but the warranty is worthwhile when you're talking about "preowned" radios. The alternative is buying directly from another ham. Hams are a pretty trustworthy group, but there are some who suffer from HDD— Honesty Deficit Disorder. If you shop the hamfest flea markets, classi-

Shop any hamfest and you'll find used equipment treasures where you least expect them!

This active QRP station has several tiny transceivers at the operator's disposal.

fied ads, on-line "swap listings" (on CompuServe, the Internet and elsewhere), do so with your eyes open. Whenever possible, try before you buy.

Shun any used transceiver that contains vacuum tubes. Tubes are quickly becoming ancient technology, and replacements are getting harder to find. As a rule of thumb, stick with solid-state gear that's no more than 10 years old. With a little savvy shopping, you'll be able to locate a good used radio with a price tag of $600 or less. And never buy a rig that doesn't include an operator's manual. It will help you avoid hours of frustration.

If $600 still has you reaching for the smelling salts, it's time to consider a *single*-band, *single*-mode transceiver. That almost always means a low-power QRP transceiver. A pre-built transceiver will set you back about $180. If you can put together an electronic kit, expect to spend between $100 and $150.

You can have a lot of fun with just a few watts of power on CW. Under the right conditions, 5 W will go just as far as 100 W. Your signal may not pin the other fellow's S meter to the wall, but it really doesn't matter. If he can understand your transmission, that's all that counts. You'll reap the benefit of using a radio that doesn't cost much, doesn't require much space, and doesn't interfere with your neighbor's TV or other electronic toys.

Amplifiers

If you feel that the RF output of your radio is too anemic, will an amplifier really help? Hmmm . . . maybe. Boosting your output from 100 W (the typical output of a multimode, multiband radio) to 1500 W *will* have a positive effect at the receiving end. But it may also have a negative effect on every electronic device in your home, your neighbor's home, *his* neighbor's home . . . you get the idea.

Amplifiers also make substantial impacts on bank accounts. A 1500-W powerhouse can set you back as much as $4000. Of course, you may also need some electrical work in your home to route a high-current ac line (usually 220 V) to your radio room. That may come at a price, too.

Am I advocating that you "say no" to HF amplifiers? Not at all. Amps have their applications, and if they didn't work they wouldn't sell as well as they do. When band conditions are mediocre or poor, an amplifier might make the difference between being heard or being just another whisper in the static. They're the favorite tools of competitive contesters and DXers who *must* be heard, no matter what.

The problem with HF amplifiers is their high cost and their tendency to cause interference. A 100-W radio by itself is capable of causing plenty of interference to televisions, telephones, stereos, VCRs, alarm systems and so on. Boost that power by a factor of 15 and guess what happens? If you don't have neighbors nearby, or if you feel competent to track down

When you want to really burn the airwaves, only an amplifier will do! A 1-kW amplifier will set you back anywhere between $1200 and $4000.

and cure the interference problems, perhaps an amp is in your future (assuming you have the cash, too).

Antennas

When you think of an antenna for the HF bands, what sort of image appears in your mind? Do you see a gleaming steel tower, majestically supporting a huge rotating beam antenna? That's the vision most of us conjure. The radio tower is an ancient icon in our hobby. It symbolizes the art and mystery of radio itself.

But ham antennas come in almost every design imaginable. Some are little more than strands of finely tuned wire. Others are thorny javelins of polished metal. Some sit atop towers. Others don't. Some look like monstrous spider webs, while others are modern-art sculptures of aluminum tubing.

There are far too many antenna designs to discuss in the few pages we have available. If you want the complete story, pick up the latest edition of *The ARRL Antenna Book*. This is the ultimate A-to-Z antenna guide, and it even includes software.

In the meantime, let's take a look at several of the most popular antennas you're likely to encounter. We'll begin with the big kids on the block . . . the beams.

HF Beams

When hams speak of beam antennas, they usually mean the venerable Yagi and quad designs (see Figure 7-1). These antennas focus your signal in a particular direction (like a flashlight). Not only do they concentrate your transmitted signal, they allow you to focus your *receive* pattern as well. For example, if your

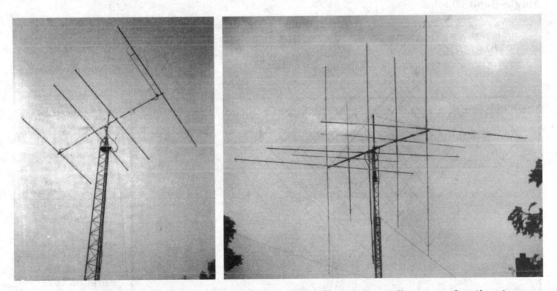

Figure 7-1—The classic HF Yagi antenna (left) is often found atop tall towers. Another tower hog is the *quad* antenna (right). This three-band HF quad is an impressive sight above any home! Both antennas require heavy-duty rotators so that they can be aimed in the desired directions.

beam is aimed west you won't hear many signals from the east (off the "back" of the beam).

The problems with HF beam antenna systems are size and cost. HF beams for the lower bands are *big* antennas. At about 43 feet in width, the longest element of a 40-meter coil-loaded Yagi is wider than the wingspan of a Piper Cherokee airplane. Even a 10-meter beam is about 18 feet across.

And a multiband (20, 15 and 10 meter) beam antenna and a 75-foot crank-up tower will set you back *at least* $2500. Then add about $500 for the antenna rotator (a beam isn't much good if you can't turn it), cables, contractor fees (to plant the tower in the ground) and so on. In the end, you'll rack up about $3000.

If you have that much cash burning a hole in your pocket, by all means throw it at a beam antenna and tower. The rewards will be tremendous. Between the signal-concentrating ability of the beam and the height advantage of the tower, you'll have the world at your fingertips. Even a beam antenna mounted on a roof tripod can make your signal an RF juggernaut. If you intend to become an avid DX chaser or con-tester, toss another $2000 into an amplifier and you'll be slugging it out with the best of them.

In truth, only a minority of hams can afford towers these days. Those who man-age to scrape together the necessary funds occasionally find themselves the targets of angry neighbors and hostile town zoning boards. (They don't appreciate the beauty of aluminum and steel like we do!)

But do you *need* a beam and a tower to enjoy Amateur Radio? This issue isn't whether they're worthwhile (they are). The question is: Are they absolutely neces-sary? The answer, thankfully for most us, is *no*.

Single-Band Dipoles

You can enjoy Amateur Radio on the HF bands with nothing more than a copper wire strung between two trees. This is the classic *dipole* antenna. It comes in several varieties, but they all function in essentially the same way.

Single-band dipoles are among the easiest antennas to build. All you need is some

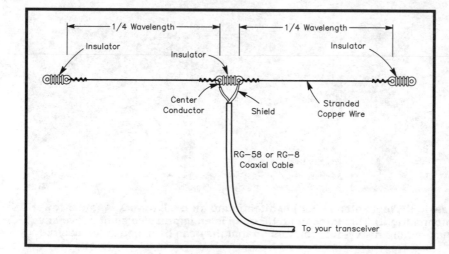

Figure 7-2—The ½-wavelength dipole is one of the easiest HF antennas to build. All you need is stranded copper wire and three insulators. You can feed a dipole with RG-58 or RG-8 coax, but you must trim the length of the antenna to achieve the lowest SWR.

stranded, noninsulated copper wire and three plastic or ceramic insulators (see Figure 7-2). A $\frac{1}{2}$-wavelength dipole is made up of two pieces of wire, each $\frac{1}{4}$-wavelength long.

Calculating the lengths of the $\frac{1}{4}$-wavelength wires is simple. Just grab a calculator and perform the following bit of division:

$$\frac{1}{4} \text{ wavelength (in feet)} = \frac{234}{\text{Frequency (in MHz)}}$$

$$\text{For example, } \frac{1}{4} \text{ wavelength at 28.4 MHz} = \frac{234}{28.4} = 8.24 \text{ feet}$$

So, to build a dipole antenna centered on this frequency you'll need two pieces of wire, each 8.24 feet long. Actually, you should add about six inches to the results of your calculations. You'll need that length margin to trim for the lowest SWR.

Join the two wires in the center with an insulator, then place insulators at both ends. Solder the center conductor of your coaxial cable to one side of the center insulator (it doesn't matter which side). Solder the shield braid of your cable to the other side. Connect ropes, nylon string or whatever to the end insulators and haul your antenna skyward. Get it as high as you can and as straight as possible. Don't hesitate to bend your dipole if that's what it takes to make it fit.

Once your dipole is safely airborne, fire up your transmitter and check the SWR at many points throughout the band. (It helps if you can plot the results on graph paper.) If you see that the SWR is getting *lower* as you move lower in frequency, your antenna is too long. Trim a couple of inches from each end and try again. On the other hand, if you see that the SWR is getting *higher* as you go lower in frequency, your antenna is too short. You'll need to *add* wire to both ends and make another series of measurements.

When you've finished trimming your dipole, you'll probably end up with an SWR of 1.5:1 or less at the center frequency, rising to 2:1 or somewhat higher at either end of the band. Don't expect a 1:1 SWR across the entire band. (Although it sometimes occurs.)

Trap Dipoles and Parallel Dipoles

For multiband applications, you'll often find the *trap* dipole (Figure 7-3) and the *parallel* dipole (Figure 7-4). Traps are tuned circuits that act somewhat like automatically switched inductors or capacitors, adding or subtracting from the length of the antenna according to the frequency of your signal. The parallel dipole uses a different approach. In the parallel design, several dipoles are joined together in the center and fed with the same cable. The dipole that radiates the RF is the one that presents an impedance that most closely matches the cable (50 Ω). That matching impedance will change according to the frequency of the signal. One dipole will offer a 50-Ω match on, say, 40 meters, while another provides the best match on 20 meters.

Obviously, these designs are somewhat more complicated than monoband dipoles, although many hams *do* choose to build their own. (See *The ARRL Antenna Book* for construction details.) If you don't have time or desire to tackle a trap or parallel dipole, you'll discover that many *QST* advertisers sell prebuilt models.

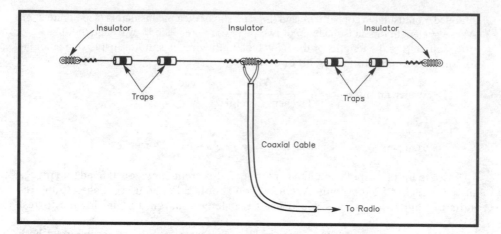

Figure 7-3—A trap dipole uses tuned circuits known as "traps" to electrically shorten or lengthen the antenna. Some elaborate trap dipoles offer coverage on many bands. You can design and build your own trap dipole, if you have enough experience. Otherwise, you're better off buying one premade.

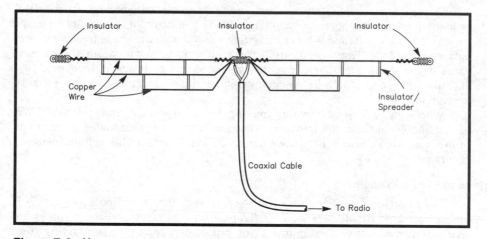

Figure 7-4—You can connect several different dipoles at the center and create the *parallel dipole* antenna. The dipoles tend to interact with each other, so building a parallel dipole usually entails *lots* of trimming and measuring. If you don't think you're up to the task, several *QST* advertisers offer prebuilt parallel dipole products.

Random-Length Dipoles

You can also enjoy multiband performance *without* traps, coils, fans or other schemes. Simply cut two equal lengths of stranded copper wire. These are going to be the two halves of your dipole antenna. Don't worry about the total length of the an-

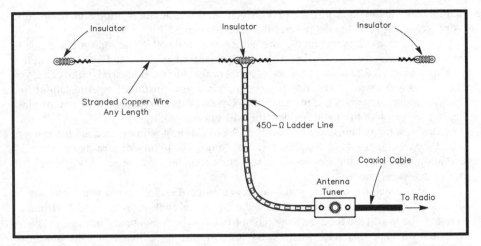

Figure 7-5—The random length dipole is simplicity itself! Just put up as much antenna wire as possible, feed it in the center with 450-Ω ladder line, and you're on the air on several bands.

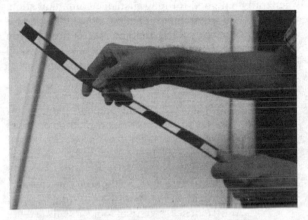

Ladder line (also called "windowed" line) is an excellent feed line for HF antennas (see Figure 7-5). It offers relatively low loss even at moderate SWRs. You need an antenna tuner with a balanced output, however, to use ladder line.

tenna. Just make it as long as possible. You won't be trimming or adding wire to this dipole.

Feed the dipole in the center with 450-Ω *ladder line* (available from most ham dealers), and buy an *antenna tuner* with a *balanced output* (see Figure 7-5). Feed the ladder line into your house, taking care to keep it from coming in contact with metal, and connect it to your tuner. Use regular coaxial cable between the antenna tuner and your radio.

You can make this antenna yourself, or buy it premade if you're short on time. A 130-foot dipole of this type should be usable on almost every HF band. Shorter versions will also work, but you may not be able to load them on every band.

Ladder line offers extremely low RF loss on HF frequencies, even when the SWR is relatively high. Just apply a signal at a low power level to the tuner and adjust the tuner controls until you achieve the lowest SWR reading. (Anything below 2:1 is fine.) You'll probably find that you need to readjust the tuner when you change frequencies. (You'll *definitely* need to readjust it when you change bands.)

You may discover that you cannot achieve an acceptable SWR on some bands, no matter how much you adjust the tuner. Even so, this antenna is almost guaranteed to work well on several bands, despite the need to retune.

So why doesn't everyone use the ladder line approach? The reason has much to do with convenience. Ladder line isn't as easy to install as coax. As I've already noted, you must keep it clear of large pieces of metal (a few inches at least). Unlike coax, you can't bend and shape ladder line to accommodate your installation. And ladder line doesn't tolerate repeated flexing as well as coaxial cable. After a year or two of playing tug 'o war with the wind, ladder line will often break.

Besides, many hams don't relish the idea of fiddling with an antenna tuner every time they change bands or frequencies. They enjoy the luxury of turning on the radio and jumping right on the air—without squinting at an antenna tuner's SWR meter and twisting several knobs.

Even with all the hassles, you can't beat a ladder-line fed dipole when it comes to sheer lack of complexity. Wire antennas fed with coaxial cable must be carefully trimmed to render the lowest SWR on each operating band. (Go back and read Chapter 2 if you've forgotten what happens to RF when the SWR gets too high.) With a ladder line dipole, no pruning is necessary. You don't even care how long it is. Simply throw it up in the air and let the tuner worry about providing a low SWR for the transceiver.

Whichever dipole you finally choose, install it as high as possible. If a horizontal dipole is too close to the ground, the lion's share of your signal will be going skyward at a steep angle. Without wading chest deep into propagation analysis, the bottom line is that a high radiation angle is generally not good for long-distance communication. Forty to 70 feet is generally considered the ideal height range, but don't lose sleep if you fall short. Raise the antenna as high as you can and change the subject when you're asked about it. You'll still make lots of contacts.

Verticals

The vertical is a popular antenna among hams who lack the space for a beam or dipole. In an electrical sense, a vertical is a dipole with half of its length buried in the ground or "mirrored" in its counterpoise system. Verticals are commonly installed at ground level, although you can also place a vertical on the roof of a building.

At first glance, a vertical looks like little more than a metal pole jutting skyward. A single-band vertical may be exactly that! However, if you look closer you'll find a network of wires snaking away in all directions from the base of the antenna. In

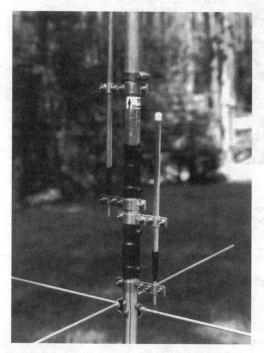

Figure 7-6—You're looking at a small section of the Cushcraft R7 multiband vertical antenna. The trap coils are wound on insulated forms, and capacitors are made from aluminum tubing. This design does not require radials.

many instances, the wires are buried a few inches beneath the soil. These are the vertical's *radials*. They provide the essential ground connection that creates the "other half" of the antenna. Multiband verticals use several traps (Figure 7-6) or similar circuits to electrically change the length of the antenna according to the frequency of the transmitted signal. (The traps are in the vertical elements, not the radials.)

Vertical antennas take little horizontal space, but they can be quite tall. Most are at least $1/4$-wavelength long at the lowest frequency. To put this in perspective, an 80-meter full-sized vertical can be over 60 feet tall! Then there is the space required by all those radial wires. You don't have to run the radials in straight lines (see Figure 7-7). In fact, you don't even have to run them underground. But you *do* need to install as many radials as possible for each band on which the antenna operates. Depending on the type of soil in your area, you may get away with a dozen radials, or you may have to install as many as 100.

Contemplate spending several days on your hands and knees pushing radial wires beneath the sod. It isn't a pretty picture, is it? That's why several antenna manufacturers developed verticals that do not use radials at all. There are questions concerning the efficiency of these antennas, but many hams swear by them (as opposed to *at* them).

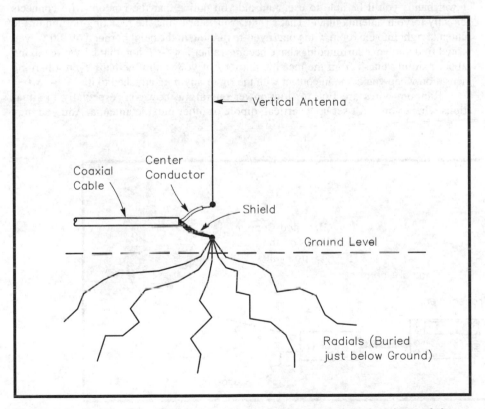

Figure 7-7—You don't have to lay down your vertical's radial wires in straight lines. Bend and twist them as much as necessary to fit your available space.

So how does the vertical stack up against the dipole when it comes to performance? If you have a generous radial system, the vertical can do at least as well as a dipole in many circumstances. Some claim that the vertical has a special advantage for DXing because it sends the RF away at a low angle to the horizon. Low radiation angles often mean longer paths as the signal bends through the ionosphere.

Without a decent radial system, however, the vertical is a poor cousin to the dipole. The old joke, "A vertical radiates equally poorly in all directions," often applies when the ground connection is lacking, such as when the soil conductivity is poor. If you can't lay down a spiderweb of radials, dipoles are often better choices.

Random Wires

A random wire is exactly that—a piece of wire that's as long as you can possibly make it. One end of the wire attaches to a tree, pole or other support, preferably at a high point. The other end connects to the random-wire connector on a suitable antenna tuner (Figure 7- 8). You apply a little RF and adjust the antenna tuner to achieve the lowest SWR. That's about all there is to it.

Random-wire antennas seem incredibly simple, don't they? The only catch is that your antenna tuner may not be able to find a match on every band. The shorter the wire, the fewer bands you'll be able to use. And did you notice that the random wire connects directly to your antenna tuner? That's right. You're bringing the radiating portion of the antenna right into the room with you. If you're running in the neighborhood of 100 W, you could find that your surroundings have become rather hot—*RF* hot, that is! We're talking about painful "bites" from the metallic portions of your radio, perhaps even a burning sensation when you come in contact with the rig or anything attached to it.

Random wires are fine for low-power operating, however, especially in situations where you can't set up a vertical, dipole or other outside antenna. And you may

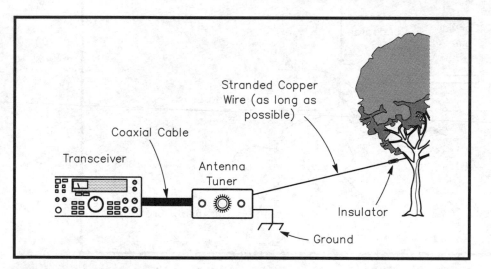

Figure 7-8—If a beam, dipole or vertical antenna doesn't work for you, consider the random end-fed wire.

be able to get away with higher power levels if your antenna tuner is connected to a good Earth ground. (A random-wire antenna needs a good ground regardless of how much power you're running.) If your radio room is in the basement or on the first floor, you may be able to use a cold water pipe or utility ground. On higher floors you'll need a *counterpoise*.

A counterpoise is simply a long, insulated wire that attaches to the ground connection on your antenna tuner. The best counterpoise is $\frac{1}{4}$-wavelength at the lowest frequency you intend to use. That's a lot of wire at, say, 3.5 MHz, but you can loop the wire around the room and hide it from view. The counterpoise acts as the other "terminal" of your antenna system, effectively balancing it from an electrical standpoint.

Indoor Antennas

So you say that you can't put up an outdoor antenna of any kind? There's hope for you yet. Antennas generally perform best when they're out in the clear, but there is no law that says you can't use an outdoor antenna *indoors*.

If you have some sort of attic in your home, apartment or condo, you're in luck. Attics are great locations for indoor antennas. For example, you can install a wire dipole in almost any attic space. Don't worry if you lack the room to run the dipole in a straight line. Bend the wires as much as necessary to make the dipole fit into the available space (see Figure 7-9).

Of course, this unorthodox installation will probably require you to spend some time trimming and tweaking the length of the antenna to achieve the lowest SWR (anything below 2:1 is fine). Not only will the antenna behave oddly because of the folding, it will probably interact with nearby electrical wiring.

Ladder-line fed dipoles are ideal for attic use—assuming that you can route the ladder line to your radio without too much metal contact. In the case of the ladder-line dipole, just make it as long as possible and stuff it into your attic any way you can. Let your antenna tuner worry about getting the best SWR out of this system.

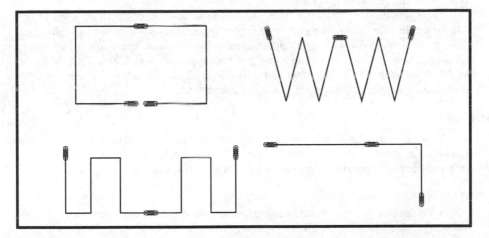

Figure 7-9—You can bend dipoles into all kinds of creative shapes to fit your yard, room, or attic.

Loop antennas like this one are ideal for indoor use. The MFJ model 1786 offers 10 to 30-MHz performance in an antenna that's only 3 feet in diameter.

Never say never! Bob Derbacher, W6REF, installed this 10-meter antenna inside his townhouse. When Bob's not using it as an antenna, it becomes a curtain rod!

For small attics you may want to consider a small tunable *loop* antenna, such as one of those manufactured by Advanced Electronic Applications or MFJ Enterprises. These incredibly small antennas offer decent performance on all but the lowest HF bands. They're basically thick circles of metal that are tuned by large, motor-driven capacitors. By adjusting a control at your radio, you change the capacitor setting and, as a result, the resonant frequency of the loop.

The same dipoles and loops that you use in your attic can also be used in any other room in your home. The same techniques apply. Keep the antenna as high off the floor as possible. (As with most antennas, the more height, the better.) For indoor operating, however, I recommend using low output power. You'll avoid RF "bites" as well as interference to VCRs, TVs and so on. Many hams have been successful operating indoor antennas with just a few watts output.

Accessories

Now that you've purchased your transceiver and antenna system, it's time to take a quick look at some station accessories. If you still have some disposable income nestled in your bank account, consider the following adornments for your operating area . . .

☐ **SWR meter**

SWR meters provide visual indications of the impedance match between your transceiver and your antenna system. They measure the magnitude of forward and reflected power, displaying the result as a *standing wave ratio* (SWR). Generally speaking, an SWR of 2:1 or less is acceptable to modern solid-state transceivers. At SWRs

This innocent-looking box is one of the most essential pieces of test equipment in your HF station: the SWR meter.

greater than 2:1, most transceivers activate a "foldback" circuit to prevent damage to the output transistors. The foldback curtails the RF output power, dropping it precipitously depending on the severity of the SWR.

Many rigs include an SWR meter. Make sure yours doesn't before you make an extra purchase! And if you intend to buy an antenna tuner (see below), you'll be pleased to know that most tuners include their own SWR meters.

Obsessive creatures that we are, most hams prefer to monitor their SWR at all times. A screwy SWR reading tells you right away that something is wrong with your antenna system. (If an overnight ice storm coats your antenna, you'll see the result as an elevated SWR the following morning.) Modern SWR meters also display your average output power, which is another nice thing to know.

So, if your radio doesn't include an SWR meter, and you don't need an antenna tuner, sink about $100 into a good standalone meter anyway. You'll be glad you did.

☐ **Antenna Tuner**

Consider this an adjustable impedance matcher between your antenna system and your transceiver. An antenna tuner doesn't really "tune" the antenna in a literal sense.

Antenna tuners come in all sizes according to how much RF power they can handle and the designs of their matching circuits. Prices range from about $80 to well over $1000. If you're using the typical 100-W transceiver, you should be able to find an adequate tuner in the $120 to $200 price range.

If you intend to use an antenna fed with ladder line, or a random-wire antenna, you *must* have an antenna tuner. For coaxial-fed antennas, however, tuners are optional. If the SWR at your operating frequency is less than 2:1 on a coax-fed antenna, you don't need a tuner. (Besides, some HF transceivers feature built-in tuners.)

Antenna tuners comes in all shapes and sizes. An economy tuner such as the MFJ-949E will match random wires, ladder-line fed dipoles, or coax-fed antennas.

The Nye Viking MB-V-A is a "high-end" antenna tuner designed to match just about anything. Although it's expensive, the tuner is built like a tank and can easily handle 1500 W.

Antenna Tuner—Do You Need One?

Buy an Antenna Tuner if . . .

. . . you want to feed your antenna with open-wire line.

Open wire line (or *ladder line*) offers extremely low loss at HF frequencies (much better than coaxial cable). One problem is that open wire line is *balanced* while your transceiver output is *unbalanced*. You need to use an antenna tuner with a built-in *balun* to form a bridge between the balanced line and the unbalanced output of your radio. A balun is a type of transformer that converts balanced feed lines to unbalanced, or vice versa. (*BAL*anced to *UN*balanced. Get it?) Most antenna tuners use 4:1 baluns, which also convert the impedance of open-wire feed lines to a value that the tuner can handle.

. . . .you want to operate your antenna on bands other than those it was designed for.

When you attempt to use, say, a 40-meter dipole on 10 meters, a big mismatch will develop along with a high SWR. By using an antenna tuner, you may be able to create a 1:1 SWR at your transceiver. (I say "may" because the mismatch can often be so great that it is beyond the capability of your tuner to handle.) The high SWR may cause substantial loss in a coaxial feed line, but at least you'll radiate *some* power at the antenna.

. . . your antenna has a narrow SWR bandwidth on some bands.

Some types of multiband antennas do not offer low SWRs from one end of each band to the other. There is usually a range—expressed in kilohertz—where an SWR below 2:1 can be achieved. For example, a multiband trap dipole may offer an SWR of 2:1 or less from 3600 to 3800 kHz. That's an SWR bandwidth of 200 kHz. If you try to operate above 3800 kHz or below 3600 kHz, you'll encounter an SWR higher than 2:1 and your radio may become displeased. With an antenna tuner, you can operate outside the SWR bandwidth and still load the full output of your radio into the antenna system.

Don't Bother With an Antenna Tuner if . . .

. . . your SWR is 2:1 or less at the frequencies you operate most often.

An SWR of 2:1 or less is not serious and does not require the assistance of an antenna tuner. Most modern rigs will tolerate a 2:1 SWR just fine. If you are using a good-quality feed line, the loss caused by an SWR of 2:1 isn't enough to worry about at HF frequencies. Many hams are obsessed with providing an absolute 1:1 SWR for their radios at all times. Apparently they also have money to burn!

. . . you're interfering with TVs, telephones and other appliances in your neighborhood.

Despite what you may have heard, an antenna tuner will not necessarily cure your interference problems. It's true that an antenna tuner may reduce the level of *harmonic radiation* (signals your radio generates in addition to the ones you want). If the interference is being caused by harmonics, a tuner may help. However, most interference is caused by RF energy that's picked up indirectly by cables or wires, or directly by the device itself. By using an antenna tuner, you'll probably radiate more energy at the antenna than you did before. That may make your interference problem worse!

Looking for Mr Goodtuner

So, you've decided that you need an antenna tuner after all. Antenna tuners come in all shapes and sizes. What features should you consider?

• A built-in SWR meter

An SWR meter of some type is a must if you want to use an antenna tuner. When adjusting your tuner, you need to keep your eye on the *reflected power* indicator. Your goal is to reduce the reflected power to zero—or at least as close as you can get. When the reflected power is zero, the SWR is 1:1 at your transceiver.

• A roller or tapped inductor

More expensive tuners feature a variable coil called a *roller inductor*. As you turn the front-panel inductor knob, the coil inside the tuner rotates. A metal wheel rolls along the coil like a train on a railroad track. As the wheel moves along the coil, the inductance increases or decreases.

Some tuners do not use roller inductors. Instead, there is a coil with wires attached at various points. On the front panel, a rotary switch selects the wires. According to how the inductor is wired in the circuit, selecting one *tap* or another varies the inductance. This is known as a *tapped inductor*.

There are advantages and disadvantages to both approaches. Roller inductors offer the best tuning performance, but they are subject to the woes of mechanical wear and tear. For example, if corrosion builds up on the wheel or the coil windings, the electrical quality of the connection will deteriorate. Roller inductors are also cumbersome to use. You may have to twist the control many times when moving from one band to another.

Tapped inductors are easy to use and free of mechanical problems (unless the switches get dirty). However, you may find that they restrict the operating range of your tuner. When you turn the switch, you select a *fixed* amount of inductance. You can't easily change it to tune a particularly difficult mismatch situation.

• A built-in balun

If you ever intend to use an open-wire feed line, buy a tuner with a 4:1 balun built-in. These baluns often dissipate quite a bit of heat, so always choose a large balun over a small one.

• Multiple antenna capability and dummy loads

Some tuners offer the ability to connect more than one antenna. This is handy in all sorts of applications. Let's say you have a vertical antenna for 10-10 meters and a wire dipole for 80 meters. You can connect both food lines to your tuner and easily switch between them.

Built-in dummy loads are convenient, but not necessary. A dummy load is a resistor (or group of resistors) that absorbs the output of your transceiver while allowing very little energy to radiate. It's used for making transmitter adjustments and other tests. If your tuner lacks a dummy load, you can purchase one separately.

A Word about Power Ratings

If your transceiver produces only 50 or 100 W of power, a 200 or 300-W tuner should do the trick, right? Well . . . yes and no. A high SWR can result in high RF voltages in the tuner. If you're trying to use your tuner in a high-SWR situation, the RF voltages at the tuner may cause an unpleasant phenomenon known as *arcing*. That's when the RF energy literally jumps the gaps between the capacitor plates or coil windings. When your tuner arcs, you'll usually hear a snapping or buzzing noise. The reflected power meter will fluctuate wildly. Interference to your TV and other devices will increase dramatically. You may even see brilliant flashes of light inside your tuner!

Arcing is obviously bad news for your tuner. It's your tuner's way of saying, "Stop! I can't handle this mismatch!" There are only two cures for arcing: reduce your output until it stops, or get a tuner with a higher power rating.

High-power tuners use large capacitors and coils. The gaps between the plates and windings are greater, making it more difficult for an arc to occur. If you can afford it, you're always better off buying a tuner with a 1.5 kW rating or better. A hefty tuner costs more, but it will serve you well in the long run.

Buy or Build?

As you comb through the advertising pages of QST, you'll see many new antenna tuners for sale. The prices are often reasonable and the quality is usually good. Keep your eyes open for used tuners, too. If a used tuner is in decent condition, it's every bit as usable as a new one.

If you like to build things, however, consider an antenna tuner as your next project. Antenna tuners are very easy to construct. You can find capacitors and coils at hamfest flea markets at very low prices. Even roller inductors—the most expensive part of a roller-inductor tuner—can be found for less than $40 if you look carefully.

Your chances of success with an antenna tuner project are excellent. You have to try pretty hard to build one poorly! Best of all, you'll have the satisfaction of using a piece of equipment that you've put together yourself. *The ARRL Handbook* offers several tuner designs you can try. Heat up your soldering iron and go to it!

In this MFJ unit, the paddle key *and* the keyer coexist. The controls on the side adjust dot-dash speed and sidetone volume. You can create "canned" CW messages and store them in memory. Just press one of the top-cover buttons to play them back.

This headset gives you total hands-free operating. The microphone is part of the headphone assembly. Swing the microphone down in front of your lips and transmit!

☐ Keys and Keyers

If you want to operate CW, you'll need a CW key. So-called "straight" keys can be yours for a minimal price in some cases, but they're tiresome to operate during long conversations (or contests). "Paddle" keys are used with electronic keyers. The keyers automatically generate the dits and dahs according to which paddle you're pressing at the time. With a paddle key and keyer at your side, you can send copious amounts of code with slight movements of your thumb and index finger. Electronic keyers are considered required equipment these days by serious contesters and DXers.

☐ Headphones and Headphone/Microphone Combos

Headphones are a blessing to any ham who lives with someone, or has crabby neighbors. Radio noises are music to our ears, but they often drive others up the nearest walls. A good set of headphones is a terrific investment for domestic tranquillity. They're also wonderful aids when working those weak stations you can barely hear. You'll be amazed at how much more you can copy with the audio output of your radio in both ears.

If you want to enjoy the luxury of hands-free voice operating, consider a headphone/microphone combination. The microphone is mounted on a small boom attached to one of the ear pieces. The boom is adjustable, allowing you to position it right in front of your lips. DXers and contesters love these gadgets! They can rack up the contacts and keep their hands free to type, write, eat, wave their fists in the air, or whatever.

☐ Logs

Although FCC logging requirements are a quaint memory, hams still enjoy keeping track of their contacts. A well-maintained log is like a diary, it allows you to step back in time and rehash some memorable conversations. Logs have more practical uses, too. If someone accuses you of interfering with their television, for example, you can consult your log and see if you were actually on the air at the time. Logs can also help you get a handle on how your antenna system is performing. Do most of your contacts seem to come from

A catchy QSL card is sure to get noticed in a pile of mediocre confirmations!

particular areas of the country? If so, perhaps your antenna radiates your signal most strongly into those areas.

Personal computers are ideal machines for logging. Check the pages of *QST* and you'll find software that maintains your log, prints QSL address labels, and does just about everything else except brew your coffee.

QSL Cards

A QSL card is written verification of a contact. It contains all the vital information such as the time of the contact, the frequency, signal reports and so forth. QSLs are primarily used by hams who are chasing various awards. They need QSLs to prove that the contacts were genuine. Even if you're not seeking certificates to hang on your wall, it's a good idea to keep a stock of cards at hand. After all, *you* may be the contact someone needs to clinch their award! You'll find plenty of QSL printers advertising in the back pages of *QST* magazine. Most will send you samples of their work for a small fee.

SCARING UP CONTACTS

When you've assembled and tested your station there is nothing left to do but get on the air. I recommend several hours of *listening* before you reach for the key or microphone. Just tune through the bands and eavesdrop on as many conversations as possible. When you finally feel comfortable with the territory, it's time to throw some RF!

CW

The best way to start a CW chat is to tune around until you hear someone calling CQ. CQ means, "I wish to contact any amateur station." In time you'll learn to recognize the sound of a CQ call. It has an unmistakable rhythm!

If you can't find anyone calling CQ, perhaps you should try it yourself. A typical CQ goes like this: CQ CQ CQ DE KD4AEK KD4AEK KD4AEK K. The letter K is an invitation for any station to reply. If there is no answer, pause for 10 or 20 seconds and repeat the call.

If you hear a CQ, wait until the ham finishes transmitting (by ending with the letter K), then call him. Make your call short, like this: K5RC K5RC DE K3YL K3YL AR (AR means "end of message").

Suppose K5RC heard someone calling him, but didn't quite catch the call because of interference (QRM) or static (QRN). Then he might come back with QRZ? DE K5RC K (Who is calling me?).

The Conversation is Underway

Most HF contacts begin with an exchange of basic information: Names, locations, equipment, signal reports and even weather reports. After that, it's up to you.

RST

Every ham wants to know how well his or her signal is reaching your location. It's a point of pride! So, the best thing you can do is to give them an honest evaluation—by using the *RST* system:

R — Readability
S — Strength
T — Tone

For readability choose a number from 1 to 5. For tone (used on CW and the digital modes only) and strength, select a number from 1 to 9.

Readability:

5 = Perfectly readable
4 = Readable with practically no difficulty
3 = Readable with considerable difficulty
2 = Barely readable, occasional words distinguishable
1 = Unreadable

Strength:

9 = Extremely strong
8 = Strong
7 = Moderately strong
6 = Good
5 = Fairly good
4 = Fair

3 = Weak
2 = Very weak
1 = Faint

Tone:

9 = Perfect
8 = Near perfect
7 = Moderately pure, just a trace of distortion
6 = Some distortion
5 = Moderate distortion
4 = Rough note
3 = Very rough note
2 = Harsh and broad
1 = Extremely harsh

It's up to you to select the numbers that best describe the signal you're hearing. An SSB signal that blows your doors off is obviously 59. But another SSB signal can be weak, yet perfectly readable. That signal might earn a 55 or even a 53.

The same ideas apply on CW and the digital modes. However, you have the tone factor to consider. In this era of modern transceivers, it's rare to hear a CW or digital signal that rates anything less than a 9 in the tone department. But if you hear a signal that bears the raspy signature of, say, a failing power supply, let the fellow know by giving him an appropriate rating.

Sometimes you'll find that you have to draw the other person into the conversation. The best way to do that is to ask questions. For example, ask what the person does for a living. She's a doctor? Okay, ask about her specialty, where she practices and more. In other words, get her to talk about herself. If you ask the right questions, the conversations will unfold on their own.

During the contact, when you want the other station to take a turn, the recommended signal is KN, meaning that you want only the contacted station to come back to you. If you don't mind someone else signing in, just K ("go") is sufficient. You *don't* need to identify yourself and the other station at the beginning and end of every transmission. That wastes time. The FCC only requires you to identify *yourself* every 10 minutes.

Ending the Conversation

When you decide to end the contact, or when the other ham expresses his/her desire to end it, don't keep talking. Briefly express your thanks: TNX QSO or TNX CHAT—and then sign out: 73 SK WA1WTB DE K5KG. If you are leaving the air, add CL to the end, right after your call sign.

SSB

To get an SSB chat off the ground, you have two choices: You can call CQ, or you can answer someone who is calling CQ.

Before calling CQ, it's important to find a frequency that appears unoccupied by any other station. This may not be easy, particularly in crowded band conditions. Listen carefully—perhaps a weak DX station is on frequency.

No matter what mode you're operating, *always listen before transmitting*. Make sure the frequency isn't being used *before* you come barging in. If, after a reasonable time, the frequency seems clear, ask if the frequency is in use, followed by your call. "Is the frequency in use? This is NY2EC." If nobody replies, you're clear to call.

Keep your CQ very short. Longwinded CQs drive most hams crazy. Besides, if no one answers, you can always call again. If you call CQ three or four times and don't get a response, try another frequency.

A typical SSB CQ goes like this:

"CQ CQ Calling CQ. This is AD5YER, Alfa-Delta-Five-Yankee-Echo-Romeo, Alfa-Delta-Five-Yankee-Echo-Romeo, calling CQ and standing by."

And if you're the caller (as opposed to being the "callee"), keep the call short. Say the call sign of the station called once or twice only, followed by your call repeated twice.

"N2EEC N2EEC, this is AB2GD, Alfa-Bravo-Two-Golf-Delta, Over."

Chewing the Rag

"Rag chewing" is ham lingo for a long, enjoyable conversation. As with CW contacts, start with the basic facts: your name, location, his signal report, and possible a brief summary

continued on page 7-30

What If No One Is Calling CQ?

You're just spent 15 minutes tuning up and down the band. You're dying to talk to someone, but you can't find anyone calling CQ. Should you pull the big switch and try again later? Not necessarily.

Stop and listen to some of the conversations in progress. With luck you'll find one that is about to end. Now listen *very* carefully. Assuming that you can hear both sides of the chat, one person is probably saying

that they have to run to the store, the post office, the bathroom or another urgent errand. Scratch that guy. He doesn't want to linger. But what about the other operator? Is he or she just signing off, in no particular hurry to be anywhere? There's your target! As soon as the conversation ends, give that person a call. About 50% of the time you'll find the other operator is more than happy to start a new round of discourse.

Questions and Answers About Lightning Protection

By Mike Tracy, KC1SX
ARRL Technical Information Services Coordinator

Q: I haven't had any lightning problems yet. Why do I need protection?

A: When most hams think of lightning protection, they immediately think about ways to protect their station equipment. Although that is certainly important, you should have far more concern for the health and welfare of yourself and your family. Each year, lightning is responsible for the deaths of over 400 people in the US. Several hundred more suffer from injuries caused by lightning, such as burns, shock and other damage to the body's more vulnerable parts.

Q: How much of a threat do I face?

A: The number of local thunderstorm days per year in this country ranges from 1 to 100, depending on where you live. If you live in a location with a single thunderstorm day, that means that you have at least one opportunity for disaster to strike. The total number of strikes per year is more than 40 million. However impressive these statistics may seem, keep in mind that they do not include all lightning strikes. Lightning can occur even without a thunderstorm—whenever and wherever there is a sufficient charge build-up.

Many things are involved in determining the likelihood of a strike at your home. A brief list includes the type of structure, the materials it's made of, the location relative to other structures and so on.

Other reasons for lightning protection include fire prevention and protection of sensitive electronic equipment. Property damage statistics indicate that lightning causes over 40 million dollars damage annually to buildings and equipment in the US.

In addition, your equipment can also be damaged by other electrical disturbances such as power line switching transients and voltage surges, as well as static build-up on outside wires and antennas.

Q: But I already have lightning protection. My station is grounded and I added a lightning arrestor to the coax.

A: Your situation is typical of many hams: a single copper rod driven into the earth as a station equipment ground and an in-line coax lightning arrestor, often mounted in the shack at the operating position. For lightning protection this sort of installation is not adequate. It may even be an invitation to disaster.

Q: So what can I do?

A: Education is the key. Lightning protection is no different from any other complex technical problem; the more you know, the better you will be at making decisions about the protection you need. In this case, however, the local library may not have the information you are looking for. While most libraries have information on lightning as a natural phenomenon, only a few will have anything on lightning protection.

A good source for this information is PolyPhaser Corporation. Although this company is in the business of manufacturing lightning-protection devices, the information they offer on installations goes far beyond product promotion. PolyPhaser's book, *The "Grounds" for Lightning and EMP Protection*, is second to none for comprehensive, easy-to-understand information on grounding systems for lightning. PolyPhaser also has a quarterly newsletter, *Striking News*, that has articles on lightning protection devices and techniques. The

February and May 1994 issues of *Striking News* contain articles on Amateur Radio station protection. Complimentary copies of these issues are available from PolyPhaser (see address below).

Q: Once I've read all about lightning, what steps should I take first to add protection to my shack?

A: The most important thing to do is to keep lightning outside of your home. This includes disconnecting your equipment from the feed lines and power sources, providing a proper station ground and adding protective devices to your installation.

As *The ARRL Antenna Book* states, "The best protection from lightning is to disconnect all antennas from equipment and disconnect all equipment from power lines." When lightning strikes, it will always try to find the shortest electrical path to ground. Unless you disconnect your station equipment, you're giving the strike a good return path through your equipment!

The easiest way to remember to do this is to disconnect your station whenever you're not using it. To prevent lightning from using your feed lines as a sneak path into your shack, disconnect them outside. If you disconnect your coax and leave it lying on the floor, lightning can jump a gap of several feet to your grounded equipment. Remember that it has already traveled quite a distance through the air. A few more feet of atmosphere won't stop it (this phenomenon is known as a "side flash").

The slick approach is to install an entrance panel for your feed lines and control cables. Place the panel ground connection on the outside of your home. Don't attach it to an inside source such as the power company ground or a cold water pipe. This panel will provide a convenient disconnect point for your equipment, as well as a place to mount feed line and control cable transient protectors.

Q: I can do that. But what about my station ground system?

A: Proper grounding is critical to lightning protection. Lightning contains energy in a wide range of frequencies (which is why you can hear "static crashes" on an AM radio when a storm approaches). You must provide a low-impedance path to ground for the energy.

A single ground rod will not suffice as a lightning ground. The basic idea is to give the strike energy a place to dissipate. Given the number of station-configuration possibilities, there are too many different ground system requirements for me to detail here. The issues of *Striking News* that I mentioned earlier contain information on grounding ham installations.

Lightning Protection Information

The ARRL Antenna Book

The ARRL Handbook (Electrical Safety chapter)

Striking News, February and May, 1994; "Ham Radio Station Protection"; PolyPhaser Corporation, PO Box 9000, Minden, NV 89423-9000

The "Grounds" for Lightning and EMP Protection; PolyPhaser Corporation Lightning Protection Code (NFPA 780-1992); National Fire Protection Association, PO Box 9101, Quincy, MA 02269-9101

LPI Installation Code (LPI 175); Lightning Protection Institute, 3365 N. Arlington Hts Rd, Suite J, Arlington Heights, IL 60004.

Installation Requirements for Lightning Protection Systems (UL 96A); Underwriters Laboratories, 333-T Pfingsten Rd, Northbrook, IL 60062

National Electrical Code (NFPA 70-1993); National Fire Protection Association

Lightning and Lightning Protection; D. W. Consultants, Inc, State Route 625, PO Box D, Gainesville, VA 22065

Enjoying QRP

Operating QRP (low power, defined by the ARRL as 5-W output or less) is a popular modus operandi of thousands of hams. The thrill of communicating at low power levels is perhaps best described as being similar to the excitement experienced during your first QSO. Interestingly enough, the level of enjoyment proportionately goes up as your power goes down

The first thing anyone contemplating a jump into the QRP sport should do is cast off the notion that you must run high power (QRO). A more enlightened attitude is needed. Consider your QRP operating as an adventure, a challenge, a unique and very personal voyage on the airwaves—riding a leaf instead of a supersonic jet. It's a gentle form of communication; think "heart and soul," not "blood and guts." Shoot down your DX prey with a pea-shooter rather than a double-barrel shotgun. A positive frame of mind will set the stage for an enjoyable time with QRP. Here are some important ground rules:

1) Listen, listen, listen.
2) Call other stations, don't call CQ.
3) Expect less-than-optimum signal reports.
4) Be persistent and patient.
5) Know when to quit.

Listen to the bands and try to figure out what the prevailing propagation is. Is the skip short or long? Who's working who? Is there much interference, static or fading (QRM, QRN or QSB, respectively) present? A quick analysis of the band conditions should be the first thing you do when sitting down for a session of QRP operating. Listening will help you decide what band to operate.

Always listen, listen, listen!

Call CQ as a last resort! Most hams prefer to answer a strong signal, which you probably will not have. You will be much better off answering a CQ. Try answering someone like this: WJ1Z DE KA1CV/QRP or WJ1Z DE KR1S/2W K. This tells WJ1Z why you aren't doing a meltdown on his headphones!

Don't be discouraged if you receive signal reports like RST 249 or RS 33. With less than 5-W output, you can't expect to be overloading the receiver front ends out there in DX-land. Have faith, for you will get more than your fair share of very respectable reports. The ultimate ego gratification of a 599 or 59 will be yours if you keep at it!

If at first you don't succeed, then try again. And again. This QRP stuff is a game of persistence, so don't give up if you don't get an answer to your call on the first try. Don't be discouraged. Make up your mind—instant success just isn't part of the plan—that's what makes it so much fun.

A very useful tool for the QRP station is the wattmeter. A commercially built unit can be found for a reasonable price. A basic single meter unit with switchable forward and reverse power is a good way to start. In time you may want to add another meter to eliminate the need to switch back and forth between forward and reverse power. Save the switch and use it to change the power range of the meter. This way you can have one range for a 5-W full scale, and the other a 1-W scale. You can calibrate this wattmeter with a VTVM (vacuum-tube voltmeter), a simple homebrew dummy load and an RF probe. Not only will you be saving some hard-earned bucks, but you will be gaining experience in designing, building, modifying and calibrating test equipment.

With QRP, your antenna is going to be much more important and instrumental in your success than if you run QRO. Running 100 W into a random wire will net you plenty of solid contacts. But when you reduce power into the same wire, your signal effectiveness will decrease, too. As a result, the old axiom of putting up the biggest antenna you can muster, as high and as in the clear as possible, means more to the QRPer than someone running 100 or 1000 W. The important thing is to optimize your antenna to your own personal circumstances. Many operators have reported amazing results using less-than-optimum antenna systems, but this is not to say you should be lax in your antenna installation. By running QRP, you are already reducing your effective radiated power (ERP); no need for a further (unintentional) reduction by cutting corners on your antenna system.

For 160 to 30 meters, dipoles generally will work well. Height is always nice, so do the best you can. Loops, end-fed wires and verticals are also used on these bands. The popular HF DX bands, 20 to 10 meters, deserve some serious thought as to rotatable gain antennas—Yagis or quads. Although this train of thought usually leads to a considerable outlay of cash, you will benefit in several ways. A 1-W signal to a 10-dB gain Yagi will give you an effective radiated output of about 10 W! That's just like having an amplifier that needs no power to run. A directional antenna is a reciprocal device as well. It is effective on both received signals and the transmitted signals. Listening to Europeans is so much more fun when you don't have to hear them along with signals from other unwanted directions.

On the bands, CW QRP activity centers around 1810, 3560, 7040 (7030 for QRP DX), 10,106, 14,060, 21,060 and 28,060 kHz. Voice operation is around 3,985, 7,285, 14,285, 21,385 and 28,885 kHz. The 10, 18 and 24-MHz bands are also hotbeds of QRP activity. Novices should check 3,710, 7,110, 21,110 and 28,110 kHz. —*Jeff Bauer, WA1MBK*

The Importance of *Zero-Beating*

By Bob Shrader, W6BNB

Very simply stated, *zero-beating* is the process of one ham tuning his transmitter to exactly the same frequency as that of a station he is receiving. The *beat note* (ie, the frequency difference) of the two stations is zero. When you listen to two hams talking on SSB, they are zero-beat if both their voices sound normal, as you listen without your touching any controls on your receiver (or transceiver). When you listen to two hams talking on CW, they are zero-beat if you hear the note of both transmitted signals at exactly the same audio pitch, again without your touching any of the controls on your receiver (or transceiver).

It is simply good operating practice to have both hams in a QSO on the same frequency, so as not to spread out and take up a wider portion of the band than is necessary. This becomes even more important in net or roundtable operation, where several CW stations who are mistuned by a few hundred hertz can cause the bandwidth in use to be perhaps two to five times that which would be used if the stations were zero-beat. Furthermore, it causes you to be continually adjusting your RIT (receiver incremental tuning) control to copy each of the different stations.

Are you beginning to see the importance of zero-beating? Well, then, let's see what's happening on the air these days.

Our HF bands are often full of signals, so we need to work carefully to squeeze the greatest number of stations possible into the limited spectrum. How? Well, we can reduce power to the minimum necessary to make the contact—which may help a bit. As a matter of fact, there is an FCC rule that requires that this be done! We can use rotary beam antennas, so as to minimize interference to stations in directions other than the "beamed" direction, which also minimizes the interference received from all the other directions. And (drum roll and cymbal crash) we should *always* accurately zero-beat the station we're talking with, so as to minimize the bandwidth used for the contact!

"How To Do It" for SSB Enthusiasts

First make sure your receiver incremental tuning (RIT) control (sometimes called OFFSET) is turned off (or, if it doesn't have an off position,

the RIT must be set to the middle of its range). With the RIT control turned off, your transmitter and receiver automatically transmit and receive on the same (carrier) frequency. To zero-beat a signal, all you have to do is to listen carefully to the voice characteristics of the signal you're tuning in. Adjust your frequency control to make the voice sound as natural as possible. Listen to all the components of the voice; you don't want to hear any strange low-pitched components or any garbled high-pitched components. Practice tuning in SSB stations with this thought in mind, and you will soon develop the preference for properly tuned signals, and a corresponding intolerance for the strange sounds of improperly tuned signals. This is to ensure that your carrier frequency is zero-beat with the other station's carrier frequency all of the time.

And then there is the older SSB equipment, where you have a separate transmitter and receiver. First, tune your receiver for the naturalness of voice that was just described. Then turn your transmitter's microphone gain down to zero and key up the transmitted signal. Tune your transmitter's frequency control until you start to hear a whistling sound, and continue tuning until the whistle goes to its lowest-frequency tone and, finally to no whistling sound at all. Viola! Your transmitter is now zero-beat with the received station. You won't interfere with other stations that might be listening to the station you are zero-beating; with SSB transmitters, the carrier frequency is suppressed and nothing is transmitted until you start talking. By the way, remember to reset your mike gain to its proper operating position after you have finished zero-beating, or you won't be transmitting when you turn on the rig and start talking!

But CW Is Another Story!

Let's start by figuring out what to do if you have a separate transmitter and receiver. Tune your receiver to the station you want to zero-beat, with the audio pitch (or tone) of the signal set to the pitch you like to copy (this varies from individual to individual, so suit yourself). Reduce your transmitter output to its lowest level (or use the *spotting* function if your transmitter has one). Now adjust your transmitter frequency so that you hear your own signal, when you tap out a few dots, at the same audio pitch as that of the other station. There you are: zero-beat! After you've done it a few times, you'll get the hang of it and can do it very quickly.

But if you're using vintage equipment, you may not have *single-signal reception*. With single-signal reception, as you tune through a CW signal, you will hear little or no signal on one side of its center frequency. Without it, you will hear the signal come into the audio band pass as a high-pitched audio note as you're tuning through, decreasing to zero, then going back up in pitch on the other side. Therefore, without single-signal reception, when you try to zero-beat you can inadvertently tune your transmitter to the "wrong" side of the other station's signal. When this happens you will be pretty far off frequency; if you are copying the other signal at a 700-Hz audio pitch, then you will be twice that amount—or 1400 Hz—away from being zero-beat. If you're that far off frequency, the other station will most likely not hear your call!

Now let's consider zero-beating with the common transceiver. At first, this might seem dead easy. Not so! Different manufacturers use different *frequency offsets* between the received and the transmitted signal, and sometimes a given manufacturer will have different offsets for different models in his product line.

Virtually all transceivers, when the RIT control is turned off, will transmit on a frequency that is between 500 and 1000 Hz different than the frequency being received; this is called the frequency offset. The concept is that if you tune your transceiver to copy a received signal at an audio pitch that is equal to the offset frequency (typically about 800 Hz), you will automatically be zero-beat. Sounds good—right? But what if you prefer to copy at an audio pitch of 500 Hz? Then when you tune your transceiver frequency to the audio pitch you prefer to copy, you will be 300 Hz (ie, 800 – 500) away from zero-beat—every time! What to do?

One of the best ways to learn to zero-beat other stations is to enlist the help of two other hams—experienced CW operators, if possible. Make some arrangement so you can all communicate either on another band, such as 2-meter simplex, or via a conference call on the telephone. In the following discussion, you will play the role of Station 1, and you will attempt to zero-beat Station 2, while Station 3 listens to critique your efforts and guide you to being zero-beat with Station 2.

The three of you will talk on your alternative frequency (2 meters or whatever), and Station 2 will transmit on some frequency, say about 7.050 MHz. Once Station 3 (the observer) has tuned in Station 2 on his receiver, he will tell you to try to zero-beat Station 2. You tune to what you think is zero-beat, and then Station 3 reports to you which direction (up in frequency or down) and about how far (1 kHz, 500 Hz, 200 Hz, or whatever) you need to move to be zero-beat. After a few such directions, Station 3 will eventually get you zero-beat with Station 2.

Now the important work begins: At this point, you need to listen carefully to the audio pitch of Station 2's signal—that's the audio note at which you *must* receive another station, in order to be zero-beat. To "remember" a specific tone is almost impossible, except for the few people with the musical talent known as "perfect pitch." You could use an external audio oscillator with an adjustable output pitch, setting its pitch to match the audio pitch you are receiving from Station 2. Then, to zero-beat another station, all you need to do is to tune the transceiver's frequency until the received signal's pitch matches your preset reference pitch from the audio oscillator.

Another possibility is to use either an audio frequency tuning control, if you have one, or an external narrowband audio filter set so that its peak frequency matches the pitch of the signal from Station 2. Using this technique, you would then tune in a received signal until the volume of the received signal peaks. (The inherent accuracy on this method is not as good as that of matching the audio pitches, but it's usually good enough to get you within 100 Hz of zero-beat.)

Well, You Get the Idea

All of this may sound complex, but don't be discouraged. A little practice will enable you to learn to zero-beat other stations accurately—a good target is always to be within 30 Hz of true zero-beat. This truly is an important concept, so work at developing your zero-beating skill; when you've mastered it, you will be a more competent operator, and the skill will increase your ability to make and hold contacts on the air.

Table 7-1

1995 Most-Wanted DXCC Countries

(Information compiled by the ARRL DXCC department)

These are the top-100 most sought-after DXCC countries. If any hams operate from one of these locations, *huge* pileups are sure to follow!

Rank	DXCC Country	Call Sign Prefix	Rank	DXCC Country	Call Sign Prefix
1	Peter 1 Is	3Y	23	Kermadec Is	ZL8
2	Bhutan	A5	24	Burma	XZ
3	Libya	5A	25	Juan de Nova	FR/J
4	Andaman	VU4	26	Mellish Reef	VK9
5	Laccadive Is	VU7	27	Ethiopia	ET
6	Heard Is	VKØ	28	SMOM	1AØ
7	Tunisia	3V	29	Trinidade	PYØT
8	Uganda	5X	30	Bouvet	3Y
9	Spratly Is	1S	31	South Sandwich Is	VP8
10	Madagascar	5R	32	Iraq	Y1
11	Yemen	7O	33	Auckland-Campbell Is	ZL9
12	Ghana	9G	34	Marion Is	ZS8
13	Tromelin	FR/T	35	Amsterdam	FT/Z
14	Bangladesh	S2	36	Tristan da	ZD9
15	Iran	EP		Cunha-Gough Is	
16	Glorioso Is	FR/G	37	Burundi	9U
17	Kingman Reef	KH5K	38	Agalega &	3B6
18	Baker, Howland Is	KH1		St Brandon Is	
19	Congo	TN	39	Qatar	A7
20	Mount Athos	SV/A	40	Laos	XW
21	Macquarie Is	VK0	41	Chad	TT
22	South Georgia Is	VP8	42	Kampuchea	XU

of your station. (How much power you're running and the kind of antenna you're using.)

Once you're beyond the preamble, the topic choice is yours. The tried and true formula for success is to get the other person to talk about themselves. Any life worth living has at least *one* interesting aspect. You may have to dig this aspect out of your palaver partner, but it's often worth the effort. If all else fails, make the following request:

"Look out the window and tell me, in detail, exactly what you see."

You'll definitely throw the other person off guard—that much is guaranteed! If they're in a room without a window, don't let them off the hook. "What would you see if you *did* have a window?"

DXing 'til Dawn

Strictly speaking, DX is any contact that you make with a ham in another country. But real DX is in the eye of the beholder. If you're new to the HF bands, DX is the long chat you just had with a fellow in Great Britain. To an HF veteran, however, DX is a five-second contact with the only amateur in the Republic of Chad.

Many hams compare DXing to fishing. You can spend minutes, even hours, drifting

Rank	DXCC Country	Call Sign Prefix	Rank	DXCC Country	Call Sign Prefix
43	Cocos-Keeling Is	VK9	72	Tokelau Is	ZK3
44	Willis Is	VK9W	73	Market Reef	OJØ
45	Crozet Is	FT/W	74	Rodriguez Is	3B9
46	Annobon Is	3CØ	75	Nauru	C2
47	Syria	YK	76	Burkina Faso	XT
48	Christmas Is	VK9	77	Egypt	SU
49	Somalia	T5	78	South Orkney Is	VP8
50	Western Sahara	SØ	79	Juan Fernandez	CEØ
51	Kerguelen	FT/X	80	Miniami Torishima	JD1
52	St Peter & Paul Rocks	PYØS	81	Rwanda	9X
53	Malpelo Is	HKØ	82	Central Kiribati	T31
54	South Sudan	STØ	83	Belau	KC6
55	Conway Reef	3D2	84	Niger	5U
56	United Arab Emirates	A6	85	Comoros Is	D6
57	Afghanistan	YA	86	Midway Is	KH4
58	Vietnam	3W	87	Guinea-Bissau	J5
59	Angola	D2	88	Togo	5V
60	Cocos Is	TI9	89	North Cook Is	ZK1
61	Benin	TY	90	Mauritania	5T
62	Sable Is	CY	91	Wallis & Futuna Is	FW
63	Aves Is	YVØ	92	Malyj Vystoskij Is	4J
64	Nepal	9N	93	Tuvalu	T2
65	Penguin Is	ZS1	94	Clipperton Is	FO/X
66	UK Sovereign Bases	ZC4	95	Niue	ZK2
67	Jan Mayen	JX	96	Navassa Is	KP1
68	Banaba (Ocean Is)	T33	97	Rotuma Is	3D2
69	Palmyra Is	KH5	98	Kure Is	KH7
70	Sao Tome	S9	99	Desecheo Is	KP5
71	St Paul Is	CY9	100	Albania	ZA

across the bands, listening to one conversation after another. Then, suddenly, there's a "tug" on your line. What's that commotion? Who are all those stations calling? You listen for a lull in the RF storm. Yes, there's the DX station. Holy cow! It's Heard Island!

Now the fish has taken the bait. Your bobber just dipped below the murky surface, but the prize is far from hooked. You have to dive into the fray and do battle for the attention of the DX station.

It could take an hour or more to land your prey, or you may not land him at all. That's the thrill of the hunt. When you hear the DX station finish with one contact, it's your turn to call. In a pileup (where many stations are calling at once) it's often best to simply send (or say) your call sign once. Now listen. Has he answered anyone? If not, send your call and listen again. As soon as you hear the DX station begin a new contact, stop calling. Yes, he might hear you in the background while he's trying to copy the other station, but you'll achieve nothing other than making him angry. He might even jot down your call sign—along with a reminder to *never* answer your call!

Prepare yourself for a jolt of giddy excitement when you hear the DX operator sending *your* call sign. Yes, he heard you! You're in the spotlight—for the next few seconds anyway.

Which is the Best Mode for DX? SSB or CW?

During periods of above-average propagation conditions, Novices and Technicians will find more DX on 10-meter voice than on either of their 15 or 10-meter CW subbands. But, as go the sunspots, so goes 10 meters. When solar activity is low, the 15-meter Novice/Technician CW subband is more productive. Novices and Technicians should not overlook their CW subbands when the sun is active either, because some choice DX occasionally shows up there. (Beginners in other countries have to study the code, too.) And, because activity in the Novice 15 and 10-meter CW subbands is lower than in the 10-meter voice band, there is less competition and a greater chance for a QSO. Further, if you are really interested in DXing, you need a higher class of license and the ability to sling out the dits and dahs.

This opinion is not just the raving of an old-time CW nut! I could fill three books with the gory details of DXpeditions and other rare ones I have called for hours on SSB, when I just couldn't break the pileup. When the DX finally got on CW, I often got through on the first call. SSB receivers use wider bandwidths by necessity and all the other strong signals calling simply overload the automatic gain control (AGC) circuit. CW allows the use of much narrower filters. Also, a good CW operator develops the ability to separate signals in his brain. This ability is no mystery, because with CW we're dealing with single tones, not a mish-mash of voices.

If you have a General class license or higher, you can operate on the other HF bands not available to Novices and Technicians. There, you should also concentrate on CW, especially if you don't have a big signal. True, some of the rarer DX stations sometimes operate in the Amateur Extra Class subbands, but a majority of those operators listen higher in the band, where General and Advanced class amateurs can transmit. If you get hooked on DXing, the first time you hear a really rare station working in the Amateur Extra Class subband, you'll be surprised how fast you can get your code speed up to snuff! Meanwhile, you can readily work more than 100 countries without going into the Amateur Extra Class subband.

What about SSB? True enough, there is plenty of DX on SSB. Almost every country I hear people fighting over on SSB, however, I have easily worked on CW (and usually on more than one band). Enough said. CW is a different means of communication and no one masters it overnight. Most of us old-timers couldn't wait to get on voice (in the days when Novices were limited to CW). Many of us found our way back to CW again, though, simply because we like it better. No matter how wild the pileup may be on CW, it just doesn't sound as bad as even a low-key pileup on SSB.

Even SSB-only operators have an incentive to get their code speed up, if only to pass the Amateur Extra Class exam. There is always something happening in the Amateur Extra Class voice subbands and you won't want to miss it.

On SSB it may go like this:

"NØSEJ you're 59." (Or he may just use the suffix of your call sign: "SEJ you're 59.")

As you choke back your joy, you reply, "JT1DHF from NØSEJ. You're also 59. Thanks and 73!"

On CW the same exchange might be:

NØSEJ 599 TU ("TU" means "to you," as in "back to you")

JT1DHF DE NØSEJ QSL 599 TNX ES 73 (Translation: JT1DHF from NØSEJ.

Why is there so much DX on CW? Most American Radio amateurs can afford a rig that works both modes, but this isn't true in many foreign countries. In many parts of the world, amateurs can't buy commercial equipment at any price and obtaining crystal filters and other parts necessary to build SSB gear is difficult, if not impossible.

You won't work these folks on any other mode. CW is harder at first; you know how to speak with your voice—now you must learn to speak with your key. I am not suggesting you give up SSB. Quite the opposite: I am hoping you won't give up CW. Being flexible is the mark of a true-blue DXer.

Getting By on CW

Let's say you decide to take the plunge and you're listening on a CW subband. Suddenly, you hear a bunch of stations sending their calls for a few seconds, spread out over several kilohertz. Then, just as suddenly, all is quiet. You tune lower in frequency and hear a weaker signal sending "5NN" ("N" is an abbreviation for "9"), but it sounds more like a machine-gun burst. There's your DX station. Who is it? Stick around, you'll hear the call sign soon enough.

Now he's calling another station, but you can only copy part of the call sign because he's going so fast. If it were your call sign he was sending, would you recognize it? If not, no sense adding to the QRM by calling him. Find another station. But for now, let's say you can follow what he's sending. There he goes again, he just sent "TU." That's his way of acknowledging the other station's

report. He's listening up the band for another call now— yes, he just let loose with another burst. Wow, you really have to concentrate!

Why not listen above him, and see if you can find the station he's working. There's someone else, sending at the same machine-gun speed, sending his call at the end of his report. Okay, back to the DX station again. A few contacts later, you've got his routine figured out and you can copy most of what he's sending.

Assuming your license allows you to transmit on this frequency, why not give him a shout? You know about how long he listens before he picks out a call. And you know whether he tunes up, down, or randomly after each contact Here's a hint: If you send your call a little slower than he's sending, he might slow down when he calls you. Sometimes a slowly sent call sign stands out in a pileup of fast callers, too.

Okay, you say, here goes "KR1S." Nope, someone else got him. Let's see, the station he called was a little lower in frequency than me. Guess I'll stand pat for one more call. Here we go again. "KR1S." Yikes! He just called me! Gosh, "5NN de KR1S TU." He sent "TU." I guess he heard me. Not exactly long winded, is he! Oh, yeah, what was his call sign? I figured it out a few minutes ago and wrote it down . . . here it is. Okay, put it in the log and don't forget the UTC time and date. Hey! I just worked a new one on CW! Gee, that wasn't too bad. Where'd I put that W1AW code-practice schedule, anyway? — *From* The DXCC Companion, *by Jim Kearman, KR1S*

Understood. You're also 599. Thanks and 73.)

That's it! You're done and he's already working someone else. Congratulations!

A "Split-Decision"

Many DX stations choose to "work split." That is, they transmit on one frequency and listen for calls on another frequency (usually a group of frequencies). For example, the DX operator may transmit on 14.195 MHz and listen for calls from

14.210 to 14.220 MHz. To snag a split-frequency DX contact you'll need a radio with two VFOs. Fortunately, most modern transceivers include this feature. The trick is to make absolutely certain that you're not transmitting on the frequency where the DX station is transmitting. You want to *listen* on that frequency!

You can usually tell when a DX station is working split. If you hear a DX station making contacts and you don't hear the other sides of the conversations, suspect a split-frequency operation. As you tune above the DX station's frequency, you'll usually find a cacophony of desperate stations—the classic *pileup*.

Experienced DX stations will often announce that they're working split. For example, on SSB you may hear the operator say, "Listening 14.210 to 14.220." On CW you might copy: UP 10, meaning that he's listening 10 kHz above his transmitting frequency.

The DXCC Award

One of the driving forces of the DX obsession is the coveted DX Century Club award, otherwise known as the DXCC. The DXCC award program is managed by the ARRL. The idea is relatively simple: Make contacts with stations in as many "official" DX countries as possible. You need 100 contacts—confirmed with QSL cards—to bag the initial DXCC certificate, and there are endorsement stickers for contacts you make beyond that point. At the top of the pile are the DX elite who've worked nearly every DXCC country in the world. They receive plaques and are admitted into the hallowed sanctum of the DXCC Honor Roll.

The definition of what makes an official DXCC country is quite different than what you might imagine. Did you know that Alaska and Hawaii are considered separate DXCC countries? Many DXCC "countries" are nothing more than tiny uninhabited islands. Groups of hams spend thousands of dollars (and even risk their lives) to mount *DXpeditions* on these god-forsaken rocks. They operate for several days or several weeks, giving hams throughout the world the opportunity to confirm that they've worked one of these rare countries. (For more than you ever wanted to know about DXCC and DXing in general, pick up a copy of *The ARRL Operating Manual*.)

The downside of the global DXCC competition is that it's nearly impossible to enjoy a real conversation with a ham if he or she is in a rare or semi-rare country. You can have fascinating, long-winded chats with Britons, Germans, French, Japanese, Russians, Australians and so on. Hams in these and other countries are so numerous that they're never "rare." But don't expect a long exchange with a ham in, say, Ghana. As soon as the word gets around that he's on the air, the pileup begins. It's hard to have a friendly chat when other folks are demanding to be heard.

This certificate is one of the most sought-after pieces of paper in the amateur world: the DXCC award.

Jumping into the Nets

Whenever you have a group of hams who meet on a particular frequency at a particular time, you have a *net*. There are nets devoted to just about every purpose you can imagine. If you're a user of Kenwood equipment, there is a Kenwood user's net. If you're into amateur satellites, you can hang out with the AMSAT nets. DX nets are devoted to arranging contacts with difficult-to-work stations. *Traffic handling* is a popular net activity. Hams meet on various frequencies to relay messages from throughout the nation or the world. And then other nets are simply groups of hams who like to meet and talk about whatever is on their minds.

A net would dissolve into chaos if it wasn't for the *net control station*, or *NCS*. He or she acts like a traffic cop at a busy intersection. The NCS "calls" the net at the appropriate time. If you happen to tune in at the start of an SSB net, you might hear something like this . . .

"Attention all amateurs. Attention all amateurs. This is KX6Y, net-control for the Klingon Language Net. This net meets every Saturday at 1400 UTC on this frequency

---

DX QSLing: Bureau or Direct?

QSL bureaus are useful if you're a DXer on a budget. Think of them as small post offices. As an ARRL member you have exclusive access to our Outgoing QSL Service. For just $3 per pound, they'll ship your QSL cards to bureaus overseas. (A pound is a *lot* of QSL cards!) The overseas bureaus, in turn, will send your cards to the DX stations.

To use the outgoing service, presort your cards according to the call-sign prefixes of the destination DX stations. Enclose an address label from your current copy of *QST* (to prove membership), along with a check or money order. Wrap the package securely and send it to:

ARRL Outgoing QSL Service
225 Main St
Newington, CT 06111

Working in reverse, the DX stations send cards to their local bureaus, who in turn ship them to *incoming* QSL bureaus here in the US. (Each call-sign district is served by an incoming QSL bureau.) If you have a self-addressed, stamped envelope (SASE) on file with the bureau that services your call district, they'll send the cards to you right away.

By the way, send your SASE to the in-coming bureau that serves the district according to your call sign—even if you've moved to another district. Let's say you lived in Alabama and were issued the call sign AA4XYZ. A few years later, you moved to Idaho (the seventh call district), but you kept your four-land call. You would still keep your envelope on file with the district-four bureau! Of course, if you switched to, say, KC7XYZ, you'd also switch to the district-seven bureau.

The bureau system is popular because it is so inexpensive. The drawback is that the system is very slow. It's common to wait a year or longer to receive a card via the bureau. That's why some hams prefer to mail their QSLs directly to DX stations. They usually include $2 for return air-mail postage and a self-addressed air-mail envelope. You'll find the addresses of many DX stations in the international edition of the *Radio Amateur's Callbook*. Many DX stations also announce their addresses several times while they're on the air. If you hear a DX operator say, "QSL via F3FRM," that means he wants you to send your card to his *QSL manager* (F3FRM in this case). So, you'd look up F3FRM's address and send your card accordingly.

Eleven-year-old Nicole, N2YJQ, enjoys her first Sweepstakes.

to exchange news and views concerning the language of the Klingon homeworld. Any stations wishing to join the net please call KX6Y now."

That's the cue for any interested ham to throw out his or her call sign. The NCS writes down each call sign in the order in which he receives it. At some point he may break in and say, "Okay, so far I have WB8IMY, N6ATQ, WR1B, KU7G and K1ZZ. Anyone else?"

Once the list is complete, the NCS will call each station in turn and ask if they have any questions or comments for the net. If he calls you, you have the option of speaking your piece or telling the NCS to skip to the next person. The NCS will also ask for new check-ins from time to time. If you didn't join the net at the beginning, that's your chance.

Traffic and other public-service nets are more tightly organized and follow stricter rules concerning who can say what . . . and when. For more information about these types of nets and how they work, see *The ARRL Operating Manual*.

Contest the Issue

Contesting is *hot* on the HF bands. How hot? Consider the fact that large numbers of hams devote incredible amounts of time, and huge sums of money, in the pursuit of top contest scores. Their radio rooms are filled with computers running the latest contest software and heavy-duty RF amplifiers. Their property is a virtual forest of copper, steel and aluminum.

Although the rules differ, the goal of every Amateur Radio contest is to contact as many stations as possible. Through a combination of technology—and skills born of rugged experience—elite contesters consistently vault to the tops of the heaps. But contesting can be fun and worthwhile for even the smallest operators. All you have to do is set *realistic* goals (see the sidebar "The Casual Contester").

The name of the game in contesting is speed. You need to make the contact, grab the necessary information, and get on to the next contact *immediately*. This is especially true when you're *running*. Running is the practice of staying on one frequency and calling "CQ contest." If you have a large station and a large signal, running is a worthwhile strategy. It's also true if you are lucky enough to live in a highly desirable location—at least as far as the contest is concerned! For example, if you took a ham vacation to operate from the Yukon during Sweepstakes, you could sit and run stations throughout the entire contest. (You'd only need to change bands as propagation conditions shifted.) Your "rare" signal from the Yukon would be a magnet for Sweepstakes participants everywhere!

A contest exchange is usually quite short. As the hard-boiled cops in the old *Dragnet* TV series used to say, "Just the facts, madam." Check the contest rules in *QST* well ahead of time and make sure you understand the exchange format. The simplest exchange is a contact number followed by a signal report. The first contact you make is 001. Your second contact is 002 and so on. Signal reports are often given as 59 or 599—even when they aren't. Yes, this probably seems less than honest, but tossing out 59/599 reports has become the contesting custom. Here is how a typical contest exchange sounds on SSB:

"CQ contest, CQ contest from WB8IMY."

"AB7AZ"

"AB7AZ from WB8IMY. You're 59, number 025." (He's my 25th contact.)

"QSL. You're also 59, number 100." (I'm his 100th contact.)

"Thanks, QRZ, WB8IMY."

Other exchanges may be more elaborate. You might have to give the last two digits of the year when you were first licensed, or your *CQ* or ITU zone number. (*The ARRL Operating Manual* contains complete *CQ* and ITU zone maps.)

Even if you never send in your logs for an official tally, you can still participate in a contest. A contact point is a contact point as far as the other contest stations are concerned. They'll accept your contacts and smile—even if they know that you won't show up in the scoring roster. The brief exchange counts toward *their* score, and that's all that matters.

Contests can also be quite valuable for the noncontester. An international DX contest will bring out all sorts of DX stations. Spend a couple of hours in this contest and you'll earn many contacts toward your DXCC award. In the same way, domestic contests are terrific for bagging your Worked All States award.

A contest is also a superb opportunity to test your antenna system. After the competition you can analyze your logs and determine which areas of the country (or the world) where your signal seems strongest (or weakest).

Contest Speak

Call used: The call sign used by this op (operator) during the contest.

Claimed score: The total estimated score (not officially accepted yet).

Class: Entry classification. (Single operator, single operator with more than one transmitter, etc.) "Assisted" indicates that a spotting network such as a *DX PacketCluster* was also used.

Dupe: More than one contact with the same station. A duplicate contact may or may not count toward your final score, depending upon the rules of the contest. For example, the contest rules might allow dupes if they take place on two different bands.

Mults: Multipliers worked during the contest. A multiplier does what it says—it "multiplies" your point totals. Multipliers are stations located in specific states, zones, or countries according to the rules of the contest.

Qs: The total number of contacts made during the contest

Run: Working many stations, one after the other, on the same frequency.

Rate: The number of contacts per hour. Many contest programs will give you a numerical readout of your rate. Some will even show this information in graph form

Search and pounce: Searching the bands for the multiplier stations you need. Searching and pouncing is common when you have difficulty "running" stations on a particular frequency.

The Casual Contester

By Glenn Swanson, KB1GW

For many, participation in Amateur Radio contests is casual in the extreme. They hear the flurry of contest activity and decide to give out a few contacts. They're in, out and done. They savor a little of the contest thrill and that's sufficient. Others spend more time and energy, but they manage to keep their cools at the same time.

To the hams who chase countries (otherwise known as *DXers*), the worldwide contests are opportunities to look for contacts with rare countries. Many of these countries seem to be more "radio active" during major competitions, thus offering a DX hunter the chance to bag a rare one. For a number of DXers, the contest is simply a means to an end. They don't care if they earn big scores or not.

One way to find your own comfort level is to set a goal. Start with, say, making 25 contacts in a particular contest. The next time out, increase your goal (try making 50 contacts this time). Most of all, don't spend a single second worrying about how others are doing.

For the first few outings, just compete against yourself. Once you've gained proficiency (and some confidence), you can set out in the pursuit of other contesters. Set your sights on topping the score of one of the lower-scoring locals (check the post-contest score listings published in the various Amateur Radio magazines, such as *QST*). Copy some of the totals of the lower-scoring stations in your area and file them in a safe place. When the same contest runs again, post the scores at your operating position. They're your targets!

During one contest I used these scores as benchmarks to see how I was doing. Before I knew it, I found myself having a great time crossing off the scores as I surpassed them one by one. "Hey! I'm doing pretty good!" It took a while, but hour after hour I was able to gauge my progress. A wide smile came with each new line drawn through the next score on my list. Did I cross out all of them? Nope—not even close, but I had fun!

As an added bonus, my contest efforts have helped me to identify station improvements that I'd like to make. Now I have some new goals—like a faster computer and better antennas for 40 and 160 meters. As my abilities (and those of my station) increase, my confidence grows.

Enjoying the Team Approach

When I began contesting I did it solo—just me and my radio. Then I heard about something called a "multi-multi" contest station. By asking some of the contesters who hang out on the local repeater, I found out that this term was used to describe a contest station that had multiple-transmitters with multiple operators—all at a single location. That sounds like a big-time operation! It wasn't long before some of the local contesters got tired of my questions and invited me to see for myself. "How'd you like to come out and operate during the next contest?" Were they kidding? That was like asking a teenager if he'd like to drive a new Ferrari. I could hardly wait!

Of course, my next emotion was fear. What if I made a fool of myself? I'd never been at the controls of a big station, especially one manned by several grizzled contest veterans. As it turned out, I didn't need to worry. I found that these folks were very helpful. In fact, they were used to helping newcomers (like me) become acquainted with the operation of the multi-multi station. I also discovered the thrill of being a part of a team effort. We each had a hand in the operation of the station and in the resulting score. It was a pleasure to operate from a well-equipped station. There's nothing like playing with top-notch radios connected to large antenna arrays.

Believe it or not, you can operate from your own station and still be part of a team. Some contests have special categories for "club competition." (The ARRL International DX Contest is one of these.) If you join a regional contest club, you can participate as part of your club's effort. For example, I belong to the Yankee Clipper Contest Club (YCCC), located here in New England. As a member I can participate in a contest from home and, after the contest, submit my score along with a notation that says it should apply to the YCCC's overall score. My score will be added to those submitted by other YCCC members who took part in the contest. Regional contest clubs encourage *all* of their members to submit a score. My score, no matter how small, will help the YCCC team. To find out if there is a regional contest club in your area, try to seek out the contesters who may frequent your local 2-meter repeater. If you have packet capability, drop a note on your local BBS, or on your nearest *DX PacketCluster*.

Station Automation Makes it Easier

Thanks to the work of proficient software authors (who are also contesters), there are several contest-logging programs available. These programs take the place of a standard paper log by keeping track of all the details (such as time, date, and frequency) of each contact. As you enter the call sign of each station you contact, the program keeps a running tally of your contest score. They'll also check for duplicate contacts that may not count toward your score. This is casual contesting at its finest!

Many software packages (along with the appropriate hardware), are capable of *automating* your station. You can control many station operations via the keyboard of a personal computer. Contest programs have become so sophisticated that a competitive contester is handicapped without them. If your radio is set up for computer control, these programs can change things like the operating frequency and mode (including "split" operations) from the keyboard of your PC. Additional hardware options allow you to use the keyboard to send (preprogrammed) Morse-code messages, handle your antenna switching tasks and even point your antenna in the proper direction. The integration of these programs into a contest station has allowed many hams to operate their equipment with just a few keystrokes. To find pricing and ordering information for these software packages, try checking the advertisements found in *QST* and in the *National Contest Journal*. You'll also discover information about station automation and contesting in general in *The ARRL Operating Manual*.

Table 7-3
Major Contests

Month	Contest	Scope
Jan	ARRL VHFSweepstakes	Primarily W/VE
Jan	CQ Worldwide 160-Meter Contest (CW)	International
Jan	ARRL RTTY Roundup	International
Feb	ARRL Novice Roundup	Novices/Techs work others
Feb	ARRL International DX Contest (CW)	W/VE stns work DX stns
Feb	CQ Worldwide 160-Meter Contest (phone)	International
Mar	ARRL International DX Contest (phone)	International
Mar	Spring RTTY Contest	International
Mar	CQ WPX Contest (phone)	International
May	Russian CQ-M Contest (phone and CW)	International
May	CQ WPX Contest (CW)	International
Jun	ARRL June VHF QSO Party	International
Jun	All Asian DX Contest (phone)	Asian stns work
Jun	ARRL Field Day	Primarily W/VE
Jul	IARU HF World Championship (phone and CW)	International
Jul	CQ VHF WPX Contest	International
Aug	Worked All Europe (CW)	EU stns work others
Sep	Worked All Europe (phone)	See above
Sep	ARRL September VHF QSO Party	International
Oct	CQ Worldwide DX Contest (phone)	International
Nov	ARRL Sweepstakes (CW)	W/VE
Nov	ARRL Sweepstakes (phone)	W/VE
Dec	ARRL 160-Meter Contest (CW)	International
Dec	ARRL 10-Meter Contest (phone and CW)	International

Exchange	For more information
Grid-square locator	Dec *QST*
W/VE; signal report and state/province; DX; signal report and country	Dec *CQ*; Contest Corral, Jan *QST*
W/VE signal report and state/provice; DX: signal report and serial number	Dec *QST*
Signal report and ARRL section	Jan *QST*
W/VE; signal report and state/province; DX: signal report and power	Dec *QST*
See above	See above
See above	Dec *QST*
UTC, signal report and consecutive QSO serial number	Contest Corral, May *QST*
Signal report and consecutive QSO serial number	Jan *CQ*; Contest Corral, Feb *QST*
Signal report and consecutive QSO serial number	Contest Corral, Apr *QST*
See above	See above
See above	May *QST*
Signal report and age	Contest Corral Jun *QST*
Transmitter "class" and ARRL	May *QST*
Signal report and ITU zone	April *QST*
Consecutive QSO serial number and call sign	Feb *CQ*; Contest Corral Jul *QST*
Signal report and consecutive QSO serial number	Contest Corral, May *QST*
See above	See above
Grid-square locator	Aug *QST*
Signal report and CQ zone	Sep *CQ*; Contest Corral, Oct *QST*
Consecutive QSO serial number, power level designator, call sign last 2 digits of the year you were licensed, ARRL section.	Oct *QST*
	Oct *QST*
W/VE; signal report and ARRL Section; DX; signal report	Nov *QST*
W/VE: signal report and state/province; DX: signal report and consecutive QSO serial number	Nov *QST*

THE HF DIGITAL MODES

Throughout this chapter we've been talking about hamming on the HF bands in terms of CW or SSB. Is that all there is? Hardly!

The next time you're tuning through the HF bands, jump over to 20 meters and listen to the segment that starts just above the CW portion of the band. If conditions are halfway decent, you'll hear all sorts of odd noises that are definitely *not* CW. Perhaps you'll stumble upon the warbling, musical signals of Baudot radioteletype, otherwise known as *RTTY* (pronounced "ritty"). If you twist the dial a bit further, you may hear a chorus of electronic crickets. These are the chirping dialogs of *AMTOR* (*AM*ateur *T*eleprinting *O*ver *R*adio) stations. Perhaps you also picked up the sounds of *PacTOR*, *G-TOR* or *CLOVER*. And what were those raspy, high-pitched bursts? Those are the unmistakable signatures of HF *packet*.

Multimode communications processors (MCPs) such as the AEA PK-900 offer RTTY, AMTOR, PACTOR, packet and many other modes in a single box.

These are the primary HF digital modes. They're called digital modes because the communication involves an exchange of data between one station and another. In the case of RTTY, for example, letters typed on a keyboard are translated into data by a computer or data terminal. Another device, usually a *multimode communications processor (MCP)*, accepts this data and converts it to whatever encoded audio tones are required. The tones are sent to the transmitter and away they go! At the receiving end the same process occurs in reverse: The tones are translated back into data and displayed as text on a computer or terminal screen (see Figure 7-10).

RTTY used to be the king of the HF digital modes, but its popularity is fading. RTTY

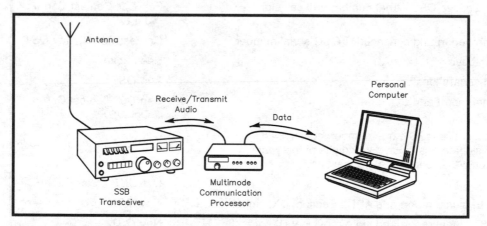

Figure 7-10—If you already own an SSB transceiver, all you need is a personal computer and a multimode processor to enter the world of HF digital communications.

is extremely easy to operate and you can use this mode with just about any radio on the market. The problem with RTTY, however, is that it doesn't include any sort of error detection. Fading and interference will reduce your nice, clean text to instant gibberish. Even so, RTTY is still favored by some DXers because you can work one station after another with reasonable speed. (RTTY is a good contest mode for the same reason.)

The so-called *burst* modes such as AMTOR, Packet, PACTOR, G-TOR and CLO-VER *do* include error detection. (They're called burst modes because they send bursts of data, rather than continuous streams of information.) Your station won't accept data from another station unless it is free of errors, although some modes have the amazing ability to "repair" damaged data automatically. The result on your screen is a continuous flow of readable text.

Which burst mode is best? I knew you were going to ask that question! Why have so many varieties if there weren't some advantages and disadvantages? To keep the length of this chapter from rivaling *War and Peace*, let's summarize . . .

- **AMTOR**: This is the first HF burst mode that hams were allowed to operate. Its error-detecting capabilities are fair, but not absolutely bullet proof. It uses a limited character set, meaning that you cannot send binary data (software and such) with AMTOR.
- **HF packet**: If band conditions are excellent and interference is virtually nonexistent, HF packet works well. Otherwise, it is one of the worst HF digital communication modes. HF packet is popular because so many *VHF* packet TNCs include HF capability.
- **PACTOR**: A fusion of AMTOR and Packet, PACTOR makes the best of both worlds. This is the most popular burst mode at the present time. PACTOR has a very robust error-detection system with the ability to repair damaged data while requiring the fewest number of repeat transmissions.
- **G-TOR**: This mode has the speed of PACTOR, but is particularly well suited for use in poor band conditions. G-TOR data throughput is typically two to three times that of PACTOR. When this book went to press, however, G-TOR was available only in the Kantronics KAM and the MFJ 1278 processors.
- **CLOVER**: In terms of speed and performance, CLOVER is the top burst mode. No other digital mode can match its throughput. CLOVER controllers are only available from HAL Communications (see the Info Guide). The controller must be plugged into an empty bus slot inside an IBM-PC or compatible computer.

You'll find the HF digital modes on every band, but 20 meters is the hot spot (see Table 7-2). Most digital operating falls into one of the following categories:

- **Bulletin Board Systems (BBSs):** Setting up a digital BBS and/or connecting to other BBSs.
- **Rag chewing:** Enjoying long, leisurely conversations.

Table 7-2
Common HF Digital Frequencies (in kHz)

1800-1840	10,140-10,150	21,070-21,100
3605-3645	14,070-14099.5	24,920-24,930
7080-7100	18,100-18,110	28,070-28,150

- **DXing:** Most DX chasing takes place using RTTY.
- **Contesting:** There are many digital contests throughout the year and they generate a *lot* of activity.

Getting started on the digital modes is easy. All you need is a multimode processor, a personal computer and an SSB transceiver. This makes digital operating a natural choice for HF-active hams who want to try something new. For the cost of a processor (about $300 to $400), they can enter a whole new world of operating pleasure. The only software you really need is a standard *terminal* program, the kind you'd use to communicate with a telephone bulletin board system. The processor manufacturers make their own software as well.

Operating the HF digital modes isn't all that difficult, but it can be confusing the first time around. Each mode has its own peculiarities. During an AMTOR, PACTOR or G-TOR conversation, for example, you must manually "turn over" the link to the other operator before he or she can send data to you. This is not true on HF packet or CLOVER. Before you jump into HF digital, I strongly recommend that you pick up a copy of *Your HF Digital Companion*. This book is available from your favorite dealer or directly from the ARRL. It gives you complete, blow-by-blow operating instructions as well as tips on setting up your HF digital station.

Other Digital and Pseudo-Digital Modes

Also available to today's digital hams are modes like ASCII (the upper- and lower-case "alphabet" used by your computer), slow-scan television (SSTV) and facsimile. Sending fax pictures on HF used to be reserved for those who happened to own expensive and bulky commercial or maritime fax transceivers. Because today's multimode processors offer black-and-white fax capabilities, sending amateur fax pictures is becoming more popular. Who knows, fax may become the next mode to explode with activity!

Info Guide

Unless otherwise indicated, all items in this section may be purchased, subject to availability, from your dealer or from the American Radio Relay League, 225 Main St, Newington, CT 06111; tel 860-594-0200. See the *Publications Catalog* in any recent issue of *QST* magazine, or call ARRL Headquarters for current pricing and ordering information.

FM AND REPEATERS

Books and Software

The ARRL Handbook for Radio Amateurs covers analog and digital electronics theory, antennas, transmitters, receivers, processing equipment and accessories.

The ARRL Operating Manual is the most complete guide to Amateur Radio operating ever published. Chapter 11 is devoted to FM and repeater operating.

The ARRL Repeater Directory is an annually published pocket-sized listing of 20,000 repeaters in the US and Canada on 29, 50-54 and 144-148, 222-225, 420-450, 902-928 and 1240 MHz and above, including FM voice, packet radio and amateur television repeaters. Operating tips, band plan charts, club listings and more.

The ARRL Electronic Repeater Directory compiled by Jerry Redington, KD6PPC, allows quick and easy access, via IBM-compatible computer, to all the information in the printed *Repeater Directory*. Computer requirements: MS-DOS, 286 or higher processor, 380 kB of system memory, 1.3 MB of hard disk space, VGA-compatible video graphics adapter and Microsoft-compatible mouse (optional). 3.5-inch disk.

The North American Repeater Atlas is written for the ham on the go who wants to keep in touch with nearby repeaters. Inside you'll find: repeater maps for every US state, every Canadian province, Mexico, Central America, and the Caribbean; street maps showing repeater frequencies for most US metropolitan areas; repeater listings for 10 meters, 2 meters, 220 MHz, 440 MHz, 900 MHz, and 1.2 GHz; and more.

PACKET RADIO AND HF DIGITAL COMMUNICATION

Books

Your Packet Companion, by Steve Ford, WB8IMY, perfect for the packet newcomer, covers everything, from assembling a station to sending mail, from packet satellites to the latest networking systems. Its straightforward writing style and clear drawings will get you on the cutting edge of digital ham radio in no time.

Practical Packet Radio by Stan Horzepa, WA1LOU, is a somewhat more technical treatment of packet communication. This books dissects the TNC and tells you how it works. You'll also learn the gritty details of packet nodes, networking and more.

Packet: Speed, More Speed and Applications is for packet enthusiasts interested in medium- to high-speed packet systems or applications that go beyond everyday messaging, BBSs and *PacketCluster*.

NOSintro: TCP/IP over Packet Radio offers a wealth of practical information, hints and tips for setting up and using the KA9Q Network Operating System (NOS) in a packet radio environment. The emphasis is on hands-on practicalities. You'll see exactly: how to install NOS on a PC, how to set up the control files, how to check out basic operations off-air, and how to use NOS commands for transferring files, logging in to remote systems, sending mail, etc.

Your HF Digital Companion, by Steve Ford, WB8IMY, takes you on a tour through the worlds of RTTY, AMTOR, HF packet, PACTOR, G-TOR and CLOVER. You'll discover how to set up your station and communicate with each of these fascinating modes.

Newsletters

Packet Status Register—Published quarterly by Tucson Amateur Packet Radio (TAPR), 8987-309 E Tanque Verde Rd, #337, Tucson, AZ 85749-9399. $15/year.

Equipment Manufacturers

Advanced Electronic Applications Inc, PO Box C2160, 2006 196th St SW, Lynnwood, WA 98036-0918; tel 206-775-7373.

Kantronics, 1202 East 23rd St, Lawrence, KS 66046-5006; tel 913-842-7745.

MFJ Enterprises, Box 494, Mississippi State, MS 39762; tel 601-323-5869.

PacComm Inc, 4413 N Hesperides St, Tampa, FL 33614-7618; tel 813-874-2980.

VHF/UHF SSB AND CW

Books

Beyond Line of Sight: A History of VHF Propagation from the Pages of QST explores the ways hams helped discover and exploit the propagation modes that allow VHF signals to travel hundreds and even thousands of miles. It's a subject all hams will find fascinating.

Radio Auroras by Charlie Newton, G2FKZ, from the RSGB, details the interesting and unpredictable world of Amateur Radio communications via auroral propagation. Presented with a European twist is information on what causes auroras, how they are forecast and how to best use them to work DX. You'll find an abundance of tables and charts.

VHF/UHF Manual, from RSGB, is must reading for the VHF and UHF enthusiast. You'll find information on the history of VHF/UHF communications, propagation, tuned circuits, receivers, transmitters, integrated equipment, filters, antennas, microwaves, space communications, and test equipment.

Organizations

Six Meter International Radio Klub (SMIRK), c/o Pat Rose W5OZI, Box 393, Junction, TX 76849. Dues are $6 per year. To apply for membership you must confirm contact with at least six SMIRK members. (Submit a log sheet indicating the date, time, frequency, call sign and membership number of each contact.)

Sidewinders on Two (SWOT), c/o Howard Hallman, WD5DGJ, 3230 Springfield, Lancaster, TX 75134. Dues are $10 per year. To apply for membership you must confirm contact with at least two SWOT members. (Submit a log sheet indicating the date, time, frequency, call sign and membership number of each contact.)

SATELLITES

Books

Satellite Experimenter's Handbook written by Martin Davidoff, K2UBC, provides the ultimate reference for the satellite operator. All active satellites are covered in detail, including telemetry formats, uplink/downlink frequencies, on-board power systems and more.

ARRL Satellite Anthology is a collection of the best satellite articles from recent issues of *QST*. A must for every satellite operator.

Newsletters

The AMSAT Journal—available from AMSAT, PO Box 27, Washington, DC 20044; tel 301-589-6062. $30/year.

OSCAR Satellite Report—available from R. Myers Communications, PO Box 17108, Fountain Hills, AZ 85269-7108. $56/year US, $62/year Canada.

Software

Software for satellite tracking, telemetry decoding and PACSAT operation is available from: AMSAT, PO Box 27, Washington, DC 20044; tel 301-589-6062. Send a self-addressed, stamped envelope and ask for their software catalog.

Equipment Manufacturers

Down East Microwave, 954 Rte 519, Frenchtown, NJ 08825; tel 908-996-3584 (downconverters, receive preamplifiers, antennas)

Hamtronics, 65-Q Moul Rd, Hilton, NY 14468; tel 716-392-9430 (downconverters)

PacComm Inc, 4413 N Hesperides St, Tampa, FL 33614-7618; tel 813-874-2980 (satellite packet TNCs)

SSB Electronic USA, 124 Cherrywood Dr, Mountaintop, PA 18707; tel 717-868-5643 (downconverters, receive preamplifiers, antennas)

AMATEUR TELEVISION

Periodicals

Amateur Television Quarterly—3 N. Court St, Crown Point, IN 46307; tel 219-662-6396

Equipment Manufacturers

Communication Concepts Inc, 508 Millstone Dr, Xenia, OH 45385; tel 513-426-8600.

Elktronics, 12536 Township Road 77, Findlay, OH 45840; tel 419-422-8206.

Micro Video Products, 1334 S Shawnee Dr, Santa Ana, CA 92704; tel 800-473-0538 or 714-957-9268.

PC Electronics, 2522 Paxson Ln, Arcadia, CA 91007-8537; tel 818-447-4565, fax 818-447-0489.

Supercircuits, 1403 Bayview Dr, Hermosa Beach, CA 90254; tel 310-372-9166.

Tactical Electronics Corp, PO Box 1743, Melbourne, FL 32902; tel 407-676-6907, fax 407-951-4630.

Wyman Research Inc, RR #1 Box 95, Waldron, IN 46182; tel 317-525-6452.

CONTACTING ARRL HEADQUARTERS VIA THE INTERNET

Awards	awards@arrl.org
Contests	contests@arrl.org
DXCC	dxcc@arrl.org
Educational Activities	ead@arrl.org
QEX magazine	qex@arrl.org
QST magazine	qst@arrl.org
Interference problems	rfi@arrl.org
Technical questions	tis@arrl.org
Exams	vec@arrl.org
W1AW—bulletins and code practice	76067.3724@compuserve.com

If you've got access to the Internet's World Wide Web, check out *ARRLWeb*. You'll find the latest W1AW bulletins, an up-to-date hamfest calendar, links to other Amateur Radio-related WWW pages and more! Just point your web browser to the URL **http://www.arrl.org/**.

ARRL AND THE ON-LINE SERVICES

The League maintains accounts on various on-line services. You can contact us via these services at the following addresses:

Compuserve — 70007,3373

America Online — HQARRL1

GEnie — ARRL

Prodigy — PTYSØ2A

. . . and don't forget the ARRL "Hiram" BBS at 860-594-0306.

Amerteur Bands and Privileges

160 meters to 23 cm

160 METERS

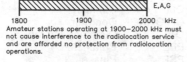

1800   1900   2000 kHz   E,A,G

Amateur stations operating at 1900–2000 kHz must not cause interference to the radiolocation service and are afforded no protection from radiolocation operations.

80 METERS

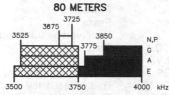

3725
3675
3525   3850
        3775   N,P
                    G
                    A
                    E
3500   3750   4000 kHz

5,167.5 kHz (SSB only) Alaska emergency use only.

40 METERS

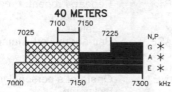

7100 7150
7025   7225   N,P
                    G   ✳
                    A   ✳
                    E   ✳
7000   7150   7300 kHz

✳Phone operation is allowed on 7075–7100 kHz in Puerto Rico; US Virgin Islands and areas of the Caribbean south of 20 degrees north latitude; and in Hawaii and areas near ITU Region 3, including Alaska.

30 METERS

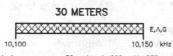

10,100   10,150 kHz   E,A,G

Maximum power on 30 meters is 200 watts PEP output. Amateurs must avoid interference to the fixed service outside the US.

20 METERS

14,025   14,150 14,225
            14,175
                            G
                            A
                            E
14,000   14,150   14,350 kHz

17 METERS

18,068   18,110   18,168 kHz   E,A,G

15 METERS

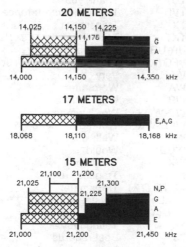

21,100 21,200
21,025   21,300
        21,225
                            N,P
                            G
                            A
                            E
21,000   21,200   21,450 kHz

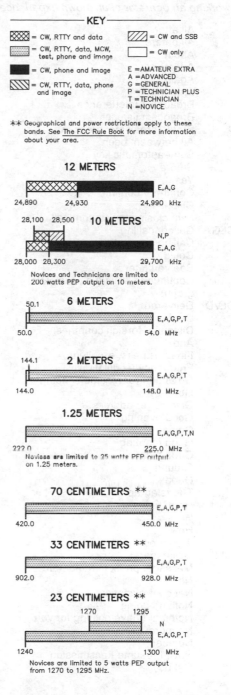

KEY

▨ = CW, RTTY and data   ▨ = CW and SSB
▦ = CW, RTTY, data, MCW,   ☐ = CW only
   test, phone and image
■ = CW, phone and image   E = AMATEUR EXTRA
                                A = ADVANCED
▨ = CW, RTTY, data, phone   G = GENERAL
   and image   P = TECHNICIAN PLUS
                                T = TECHNICIAN
                                N = NOVICE

✳✳ Geographical and power restrictions apply to these bands. See The FCC Rule Book for more information about your area.

12 METERS

24,890   24,930   24,990 kHz   E,A,G

10 METERS

28,100 28,500
28,000 28,300   29,700 kHz   N,P
                                     E,A,G

Novices and Technicians are limited to 200 watts PEP output on 10 meters.

6 METERS

50.1
50.0   54.0 MHz   E,A,G,P,T

2 METERS

144.1
144.0   148.0 MHz   E,A,G,P,T

1.25 METERS

222.0   225.0 MHz   E,A,G,P,T,N

Novices are limited to 25 watts PEP output on 1.25 meters.

70 CENTIMETERS ✳✳

420.0   450.0 MHz   E,A,G,P,T

33 CENTIMETERS ✳✳

902.0   928.0 MHz   E,A,G,P,T

23 CENTIMETERS ✳✳

1270   1295   N
1240   1300 MHz   E,A,G,P,T

Novices are limited to 5 watts PEP output from 1270 to 1295 MHz.

Some Abbreviations for CW Work

Some Abbreviations help to cut down unnecessary transmission; do not abbreviate unnecessarily when working an operator of unknown experience.

AA	All after	OC	Old chap
AB	All before	OM	Old man
ABT	About	OP-OPR	Operator
ADR	Address	OT	Old timer, old top
AGN	Again	PBL	Preamble
ANT	Antenna	PSE	Please
BCI	Broadcast Interference	PWR	Power
BCL	Broadcast listener	PX	Press
BK	Break, break me; break-in	R	Received as transmitted; are
BN	All between; been	RCD	Received
BUG	Semi-automatic key	RCVR (RX)	Receiver
B4	Before	REF	Refer to; referring to; reference
C	Yes	RFI	Radio frequency interference
CFM	Confirm; I confirm	RIG	Station equipment
CK	Check	RPT	Repeat; I repeat; report
CL	I am closing my station; call	RTTY	Radioteletype
CLD-CLG	Called; calling	RX	Receiver
CQ	Calling any station	SASE	Self-addressed, stamped envelope
CUD	Could	SED	Said
CUL	See you later	SIG	Signature; signal
CW	Continuous wave (i.e., radiotelegraph)	SINE	Operator's personal initials or nickname
DLD-DLVD	Delivered	SKED	Schedule
DR	Dear	SRI	Sorry
DX	Distance, foreign countries	SSB	Single sideband
ES	And,	SVC	Service; prefix to service message
FB	Fine business, excellent	T	Zero
FM	Frequency modulation	TFC	Traffic
GA	Go ahead (or resume sending)	TMW	Tomorrow
GB	Good-by	TNX-TKS	Thanks
GBA	Give better address	TT	That
GE	Good evening	TU	Thank you
GG	Going	TVI	Television interference
GM	Good morning	TX	Transmitter
GN	Good night	TXT	Text
GND	Ground	UR-URS	Your; you're; yours
GUD	Good	VFO	Variable-frequency oscillator
HI	The telegraphic laugh	VY	Very
HR	Here, hear	WA	Word after
HV	Have	WB	Word before
HW	How	WD-WDS	Word; words
LID	A poor operator	WKD-WKG	Worked; working
MA, MILS	Milliamperes	WL	Well; will
MSG	Message; prefix to radiogram	WUD	Would
N	No	WX	Weather
NCS	Net control station	XCVR	Transceiver
ND	Nothing doing	XMTR (TX)	Transmitter
NIL	Nothing; I have nothing for you	XTAL	Crystal
NM	No more	XYL (YF)	Wife
NR	Number	YL	Young lady
NW	Now; I resume transmission	73	Best regards
OB	Old boy	88	Love and kisses

Standard ITU Phonetics

A—Alfa (**AL** FAH)
B—Bravo (**BRAH** VOH)
C—Charlie (**CHAR** LEE) or (**SHAR** LEE)
D—Delta (**DELL** TAH)
E—Echo (**ECK** OH)
F—Foxtrot (**FOKS** TROT)
G—Golf (GOLF)
H—Hotel (HOH **TELL**)
I—India (**IN** DEE AH)
J—Juliett (**JEW** LEE ETT)
K—Kilo (**KEY** LOH)
L—Lima (**LEE** MAH)
M—Mike (MIKE)

N—November (NO **VEM** BER)
O—Oscar (**OSS** CAH)
P—Papa (PAH **PAH**)
Q—Quebec (KEH **BECK**)
R—Romeo (**ROW** ME OH)
S—Sierra (SEE **AIR** RAH)
T—Tango (**TANG** GO)
U—Uniform (**YOU** NEE FORM) or (OO NEE FORM)
V—Victor (**VIK** TAH)
W—Whiskey (**WISS** KEY)
X—X-RAY (**ECKS** RAY)
Y—Yankee (**YANG** KEY)
Z—Zulu (**ZOO** LOO)

Note: The **boldfaced** syllables are emphasized. The pronunciations shown in this table were designed for those who speak any of the international languages. The pronunciations given for "Oscar" and "Victor" may seem awkward to English-speaking people in the US.

Q Signals

These Q signals are the ones used most often on the air. (Q abbreviations take the form of questions only when they are sent followed by a question mark.)

QRG Will you tell me my exact frequency (or that of ___)? Your exact frequency (or that of ___) is ___ kHz.
QRL Are you busy? I am busy (or I am busy with ___). Please do not interfere.
QRM Is my transmission being interfered with? Your transmission is being interfered with ___ (1. Nil; 2. Slightly; 3. Moderately; 4. Severely; 5. Extremely.)
QRN Are you troubled by static? I am troubled by static ___. (1-5 as under QRM.)
QRO Shall I increase power? Increase power.
QRP Shall I decrease power? Decrease power.
QRQ Shall I send faster? Send faster (___ WPM).
QRS Shall I send more slowly? Send more slowly (___ WPM).
QRT Shall I stop sending? Stop sending.
QRU Have you anything for me? I have nothing for you.
QRV Are you ready? I am ready.
QRX When will you call me again? I will call you again at ___ hours (on ___ kHz).
QRZ Who is calling me? You are being called by ___ (on ___ kHz).
QSB Are my signals fading? Your signals are fading.
QSK Can you hear me between your signals and if so can I break in on your transmission? I can hear you between signals, break in on my transmission.
QSL Can you acknowledge receipt (of a message or transmission)? I am acknowledging receipt.
QSN Did you hear me (or ___) on ___ kHz? I did hear you (or ___) on ___ kHz.
QSO Can you communicate with ___ direct or by relay? I can communicate with ___ direct (or relay through ___).
QSP Will you relay to ___? I will relay to ___.
QST General call preceding a message addressed to all amateurs and ARRL members. This is in effect "CQ ARRL."
QSX Will you listen to ___ on ___ kHz? I am listening to ___ on ___ kHz.
QSY Shall I change to transmission on another frequency? Change to transmission on another frequency (or on ___ kHz).
QTB Do you agree with my counting of words? I do not agree with your counting of words. I will repeat the first letter or digit of each word or group.
QTC How many messages have you to send? I have ___ messages for you (or for ___).
QTH What is your location? My location is ___.
QTR What is the correct time? The time is ___.

W1AW schedule

Pacific	Mtn	Cent	East	Sun	Mon	Tue	Wed	Thu	Fri	Sat
6 am	7 am	8 am	9 am			Fast Code	Slow Code	Fast Code	Slow Code	
7 am	8 am	9 am	10 am			Code Bulletin				
8 am	9 am	10 am	11 am			Teleprinter Bulletin				
9 am	10 am	11 am	noon							
10 am	11 am	noon	1 pm			Visiting Operator Time				
11 am	noon	1 pm	2 pm							
noon	1 pm	2 pm	3 pm							
1 pm	2 pm	3 pm	4 pm	Slow Code	Fast Code	Slow Code	Fast Code	Slow Code	Fast Code	Slow Code
2 pm	3 pm	4 pm	5 pm	Code Bulletin						
3 pm	4 pm	5 pm	6 pm	Teleprinter Bulletin						
4 pm	5 pm	6 pm	7 pm	Fast Code	Slow Code	Fast Code	Slow Code	Fast Code	Slow Code	Fast Code
5 pm	6 pm	7 pm	8 pm	Code Bulletin						
6 pm	7 pm	8 pm	9 pm	Teleprinter Bulletin						
6^{45} pm	7^{45} pm	8^{45} pm	9^{45} pm	Voice Bulletin						
7 pm	8 pm	9 pm	10 pm	Slow Code	Fast Code	Slow Code	Fast Code	Slow Code	Fast Code	Slow Code
8 pm	9 pm	10 pm	11 pm	Code Bulletin						
9 pm	10 pm	11 pm	Mdnte	Teleprinter Bulletin						
9^{45} pm	10^{45} pm	11^{45} pm	12^{45} am	Voice Bulletin						

W1AW's schedule is at the same local time throughout the year. The schedule according to your local time will change if your local time does not have seasonal adjustments that are made at the same time as North American time changes between standard time and daylight time. From the first Sunday in April to the last Sunday in October, UTC = Eastern Time + 4 hours. For the rest of the year, UTC = Eastern Time + 5 hours.

• *Morse code transmissions:*

Frequencies are 1.818, 3.5815, 7.0475, 14.0475, 18.0975, 21.0675, 28.0675 and 147.555 MHz.
Slow Code = practice sent at 5, 7^{1}/$_{2}$, 10, 13 and 15 wpm.
Fast Code = practice sent at 35, 30, 25, 20, 15, 13 and 10 wpm.
Code practice text is from the pages of *QST*. The source is given at the beginning of each practice session and alternate speeds within each session. For example, "Text is from July 1992 *QST*, pages 9 and 81," indicates that the plain text is from the article on page 9 and mixed number/letter groups are from page 81. Code bulletins are sent at 18 wpm.
W1AW qualifying runs are sent on the same frequencies as the Morse code transmissions. West Coast qualifying runs are transmitted on approximately 3.590 MHz by W6OWP, with W6ZRJ and AB6YR as alternates. At the beginning of each code practice session, the schedule for the next qualifying run is presented. Underline one minute of the highest speed you copied, certify that your

copy was made without aid, and send it to ARRL for grading. Please include your name, call sign (if any) and complete mailing address. Send a 9×12-inch SASE for a certificate, or a business-size SASE for an endorsement.

• *Teleprinter transmissions:*

Frequencies are 3.625, 7.095, 14.095, 18.1025, 21.095, 28.095 and 147.555 MHz.

Bulletins are sent at 45.45-baud Baudot and 100-baud AMTOR, FEC Mode B. 110-baud ASCII will be sent only as time allows.

On Tuesdays and Saturdays at 6:30 PM Eastern Time, Keplerian elements for many amateur satellites are sent on the regular teleprinter frequencies.

• *Voice transmissions:*

Frequencies are 1.855, 3.99, 7.29, 14.29, 18.16, 21.39, 28.59 and 147.555 MHz.

• *Miscellanea:*

On Fridays, UTC, a DX bulletin replaces the regular bulletins.

W1AW is open to visitors during normal operating hours: from 1 PM until 1 AM on Mondays, 9 AM until 1 AM Tuesday through Friday, from 1 PM to 1 AM on Saturdays, and from 3:30 PM to 1 AM on Sundays. FCC licensed amateurs may operate the station from 1 to 4 PM Monday through Saturday. Be sure to bring your current FCC amateur license or a photocopy.

In a communication emergency, monitor W1AW for special bulletins as follows: voice on the hour, teleprinter at 15 minutes past the hour, and CW on the half hour.

Headquarters and W1AW are closed on New Year's Day, President's Day, Good Friday, Memorial Day, Independence Day, Labor Day, Thanksgiving and the following Friday, and Christmas Day. On the first Thursday of September, Headquarters and W1AW will be closed during the afternoon.

Voluntary HF Band Plans for Considerate US Operators

Voluntary band plans result from mutual agreement among amateurs. They add to, and are subject to, frequency allocations and subbands defined by FCC regulations.

Band (MHz)	CW	RTTY	Phone	SSTV[†]	Fax[††]	Packet	AM
1.8	1.800-1.840 1.830-1.840 DX	1.800-1.840 1.830-1.840 DX	1.840-2.000 1.840-1.850 DX				
3.5	3.500-3.580	3.580-3.620 3.590 DX	3.750-4.000 3.790-3.800 DX	3.845		3.620-3.635 3.5943[†††](Int) 3.6073[†††](NA)	3.885
7.0	7.000-7.080		7.150-7.300	7.171	7.245	7.080-7.100 7.0383[†††](Int) 7.0913[†††] (NA)	7.290
10.1	10.100-10.130					10.140-10.150 10.1453[†††](Int) 10.1473[†††](NA)	
14	14.000-14.070 14.100 (b)		14.150-14.350	14.230	14.245	14.095-14.0995 14.1005-14.112 14.1023[†††](Int) 14.1083[†††](NA)	14.286
18	18.068-18.100	18.100-18.105	18.110-18.168			18.105-18.110 18.1063[†††] 18.1083[†††]	
21	21.000-21.070	21.070-21.090	21.200-21.450	21.340	21.345	21.090-21.100 21.0963[†††] 21.0983[†††]	
24.8	24.890-24.920	24.920-24.925	24.930-24.990			24.925-24.930 24.9263[†††] 24.9283[†††]	
28	28.000-28.070 28.200-28.300 (b⁺)	28.070-28.120	28.300-29.300 29.300-29.510 (satellite downlinks) 29.520-29.680 (repeaters) 29.600 (FM simplex calling)	28.680	28.945	28.120-28.189 28.1023[†††] 28.1043[†††]	29.000-29.200

Notes

[†]   Establish contact on voice or CW first, then move off frequency to send images. A "gentlemen's agreement" calls for SSTV operation as close to the calling frequencies as possible.

[††]   Always establish voice or CW contact before sending fax.

[†††]   Automatic message forwarding: Int - Intercontinental; NA = North America

(b) = beacons

(b⁺) Domestically, beacons can be operating in this segment. Internationally, beacons can be found in the 28.190-28.225 segment.

Abbreviations List

A

a—atto (prefix for 10^{-18})
A—ampere (unit of electrical current)
ac—alternating current
ACC—Affiliated Club Coordinator
ACSSB—amplitude-compandored single sideband
A/D—analog-to-digital
ADC—analog-to-digital converter
AF—audio frequency
AFC—automatic frequency control
AFSK—audio frequency-shift keying
AGC—automatic gain control
A/h—ampere hour
AIRS—ARRL Interference Recording System
ALC—automatic level control
AM—amplitude modulation
AMRAD—Amateur Radio Research and Development Corp
AMSAT—Radio Amateur Satellite Corp
AMTOR—Amateur Teleprinting Over Radio
ANT—antenna
ARA—Amateur Radio Association
ARC—Amateur Radio Club
ARES—Amateur Radio Emergency Service
ARQ—Automatic repeat request
ARRL—American Radio Relay League
ARS—Amateur Radio Society (station)
ASCII—American National Standard Code for Information Interchange
ASSC—Amateur Satellite Service Council
ATV—amateur television
AVC—automatic volume control
AWG—American wire gauge
az-el—azimuth-elevation

B

B—bel; blower; susceptance; flux density (inductors)
balun—balanced to unbalanced (transformer)
BC—broadcast
BCD—binary coded decimal
BCI—broadcast interference
Bd—baud (bit/s in single-channel binary data transmission)
BER—bit error rate
BFO—beat-frequency oscillator
bit—binary digit
bit/s—bits per second
BM—Bulletin Manager
BPF—band-pass filter
BPL—Brass Pounders League
BT—battery
BW—bandwidth

C

c—centi (prefix for 10^{-2})
C—coulomb (quantity of electric charge); capacitor
CAC—Contest Advisory Committee
CATVI—cable television interference
CB—Citizens Band (radio)
CBBS—computer bulletin-board service
CBMS—computer-based message system
CCIR—International Research Consultative Committee
CCITT—International Telegraph and Telephone Consultative Committee
CCTV—closed-circuit television
CCW—coherent CW
ccw—counterclockwise
CD—civil defense
cm—centimeter
CMOS—complimentary-symmetry metal-oxide semiconductor
coax—coaxial cable
COR—carrier-operated relay
CP—code proficiency (award)
CPU—central processing unit
CRT—cathode ray tube
CT—center tap
CTCSS—continuous tone-coded squelch system
cw—clockwise
CW—continuous wave

D

d—deci (prefix for 10^{-1})
D—diode
da—deca (prefix for 10)
D/A—digital-to-analog
DAC—digital-to-analog converter
dB—decibel (0.1 bel)
dBi—decibels above (or below) isotropic antenna
dBm—decibels above (or below) one milliwatt
DBM—doubly balanced mixer
dBV—decibels above/below 1 V (in video, relative to 1 V P-P)
dBW—decibels above/below 1 W
dc—direct current
D-C—direct conversion

DDS—direct digital synthesis
DEC—District Emergency Coordinator
deg—degree
DET—detector
DF—direction finding; direction finder
DIP—dual in-line package
DMM—digital multimeter
DPDT—double-pole double-throw (switch)
DPSK—differential phase-shift keying
DPST—double-pole single-throw (switch)
DS—direct sequence (spread spectrum); display
DSB—double sideband
DSP—digital signal processing
DTMF—dual-tone multifrequency
DVM—digital voltmeter
DX—long distance; duplex
DXAC—DX Advisory Committee
DXCC—DX Century Club

E

e—base of natural logarithms (2.71828)
E—Voltage
EA—Educational Advisor
EC—Emergency Coordinator
ECAC—Emergency Communications Advisory
    Committee
ECL—emitter-coupled logic
EHF—extremely high frequency (30-300 GHz)
EIA—Electronic Industries Assn
EIRP—effective isotropic radiated power
ELF—extremely low frequency
ELT—emergency locator transmitter
EMC—electromagnetic compatibility
EME—earth-moon-earth (moonbounce)
EMF—electromotive force
EMI—electromagnetic interference
EMP—electromagnetic pulse
EOC—emergency operations center
EPROM—erasable programmable read only
    memory

F

f—femto (prefix for 10^{-15}); frequency
F—farad (capacitance unit); fuse
fax—facsimile
FCC—Federal Communications Commission
FD—Field Day
FEMA—Federal Emergency Management Agency
FET—field-effect transistor
FFT—fast Fourier transform
FL—filter
FM—frequency modulation

FMTV—frequency-modulated television
FSK—frequency-shift keying
FSTV—fast-scan (real-time) television
ft—foot (unit of length)

G

g—gram (unit of mass)
G—giga (prefix for 10^9); conductance
GaAs—gallium arsenide
GDO—grid- or gate-dip oscillator
GHz—gigahertz (10^{9} Hz)
GND—ground

H

h—hecto (prefix for 10^2)
H—henry (unit of inductance)
HF—high frequency (3-30 MHz)
HFO—high-frequency oscillator; heterodyne
    frequency oscillator
HPF—highest probable frequency; high-pass
    filter
Hz—hertz (unit of frequency, 1 cycle/s)

I

I—current, indicating lamp
IARU—International Amateur Radio Union
IC—integrated circuit
ID—identification; inside diameter
IEEE—Institute of Electrical and Electronics
    Engineers
IF—intermediate frequency
IMD—intermodulation distortion
in.—inch (unit of length)
in./s—inch per second (unit of velocity)
I/O—input/output
IRC—international reply coupon
ISB—independent sideband
ITF—Interference Task Force
ITU—International Telecommunication Union

J

j—operator for complex notation, as for reactive
    component of an impedance (+*j* inductive; –*j*
    capacitive)
J—joule (kg m²/s²) (energy or work unit); jack
JFET—junction field-effect transistor

K

k—kilo (prefix for 10^3); Boltzmann's constant
    (1.38×10^{-23} J/K)
K—kelvin (used without degree symbol)
    (absolute temperature scale); relay
kBd—1000 bauds

kbit—1024 bits
kbit/s—1024 bits per second
kbyte—1024 bytes
kg—kilogram
kHz—kilohertz
km—kilometer
kV—kilovolt
kW—kilowatt
kΩ—kilohm

L

l—liter (liquid volume)
L—lambert; inductor
lb—pound (force unit)
LC—inductance-capacitance
LCD—liquid crystal display
LED—light-emitting diode
LF—low frequency (30-300 kHz)
LHC—left-hand circular (polarization)
LO—local oscillator; League Official
LP—log periodic
LS—loudspeaker
lsb—least significant bit
LSB—lower sideband
LSI—large-scale integration
LUF—lowest usable frequency

M

m—meter (length); milli (prefix for 10^{-3})
M—mega (prefix for 10^6); meter (instrument)
mA—milliampere
mAh—milliamperehour
MCP—multimode communications processor
MDS—Multipoint Distribution Service; minimum discernible (or detectable) signal
MF—medium frequency (300-3000 kHz)
mH—millihenry
MHz—megahertz
mi—mile, statute (unit of length)
mi/h—mile per hour
mi/s—mile per second
mic—microphone
min—minute (time)
MIX—mixer
mm—millimeter
MOD—modulator
modem—modulator/demodulator
MOS—metal-oxide semiconductor
MOSFET—metal-oxide semiconductor field-effect transistor
MS—meteor scatter
ms—millisecond

m/s—meters per second
msb—most-significant bit
MSI—medium-scale integration
MSK—minimum-shift keying
MSO—message storage operation
MUF—maximum usable frequency
mV—millivolt
mW—milliwatt
MΩ—megohm

N

n—nano (prefix for 10^{-9}); number of turns (inductors)
NBFM—narrow-band frequency modulation
NC—no connection; normally closed
NCS—net-control station; National Communications System
nF—nanofarad
NF—noise figure
nH—nanohenry
NiCd—nickel cadmium
NM—Net Manager
NMOS—N-channel metal-oxide silicon
NO—normally open
NPN—negative-positive-negative (transistor)
NPRM—Notice of Proposed Rule Making (FCC)
NR—Novice Roundup (contest)
ns—nanosecond
NTS—National Traffic System

O

OBS—Official Bulletin Station
OD—outside diameter
OES—Official Emergency Station
OO—Official Observer
op amp—operational amplifier
ORS—Official Relay Station
OSC—oscillator
OSCAR—Orbiting Satellite Carrying Amateur Radio
OTC—Old Timer's Club
OTS—Official Traffic Station
oz—ounce (force unit, 1/16 pound)

P

p—pico (prefix for 10^{-12})
P—power; plug
PA—power amplifier
PacTOR—digital mode combining aspects of packet and AMTOR
PAM—pulse-amplitude modulation
PBS—packet bulletin-board system

PC—printed circuit
P_D—power dissipation
PEP—peak envelope power
PEV—peak envelope voltage
pF—picofarad
pH—picohenry
PIA—Public Information Assistant
PIN—positive-intrinsic-negative (semi-
    conductor)
PIO—Public Information Officer
PIV—peak inverse voltage
PLL—phase-locked loop
PM—phase modulation
PMOS—P-channel (metal-oxide semiconductor)
PNP—positive-negative positive (transistor)
pot—potentiometer
P-P—peak to peak
ppd—postpaid
PRAC—Public Relations Advisory Committee
PROM—programmable read-only memory
PSHR—Public Service Honor Roll
PTO—permeability-tuned oscillator
PTT—push to talk

Q

Q—figure of merit (tuned circuit); transistor
QRP—low power (less than 5-W output)

R

R—resistor
RACES—Radio Amateur Civil Emergency
    Service
RAM—random-access memory
RC—resistance-capacitance
R/C—radio control
RCC—Rag Chewer's Club
RDF—radio direction finding
RF—radio frequency
RFC—radio-frequency choke
RFI—radio-frequency interference
RHC—right-hand circular (polarization)
RIT—receiver incremental tuning
RLC—resistance-inductance-capacitance
RM—rule making (number assigned to petition)
r/min—revolutions per minute
RMS—root mean square
ROM—read-only memory
r/s—revolutions per second
RST—readability-strength-tone (CW signal
    report)
RTTY—radioteletype
RX—receiver, receiving

S

s—second (time)
S—siemens (unit of conductance; switch
SASE—self-addressed stamped envelope
SCF—switched capacitor filter
SCR—silicon controlled rectifier
SEC—Section Emergency Coordinator
SET—Simulated Emergency Test
SGL—State Government Liaison
SHF—super-high frequency (3-30 GHz)
SM—Section Manager; silver mica
    (capacitor)
S/N—signal-to-noise ratio
SPDT—single pole double-throw (switch)
SPST—single-pole single-throw (switch)
SS—Sweepstakes; spread spectrum
SSB—single sideband
SSC—Special Service Club
SSI—small-scale integration
SSTV—slow-scan television
STM—Section Traffic Manager
SX—simplex
sync—synchronous, synchronizing
SWL—shortwave listener
SWR—standing-wave ratio

T

T—tera (prefix for 10^{12}); transformer
TA—Technical Advisor
TC—Technical Coordinator
TCC—Transcontinental Corps (NTS)
TCP/IP—Transmission Control Protocol/Internet
    Protocol
TD—Technical Department (ARRL HQ)
tfc—traffic
TNC—terminal node controller (packet radio)
TR—transmit/receive
TS—Technical Specialist
TTL—transistor-transistor logic
TTY—teletypewriter
TU—terminal unit
TV—television
TVI—television interference
TX—transmitter, transmitting

U

U—integrated circuit
UHF—ultra-high frequency (300 MHz to
    3 GHz)
USB—upper sideband
UTC—Coordinated Universal Time
UV—ultraviolet

V

V—volt; vacuum tube
VCO—voltage-controlled oscillator
VCR—video cassette recorder
VDT—video-display terminal
VE—Volunteer Examiner
VEC—Volunteer Examiner Coordinator
VFO—variable-frequency oscillator
VHF—very-high frequency (30-300 MHz)
VLF—very-low frequency (3-30 kHz)
VLSI—very-large-scale integration
VMOS—V-topology metal-oxide semiconductor
VOM—volt-ohm meter
VOX—voice operated switch
VR—voltage regulator
VRAC—VHF Repeater Advisory Committee
VSWR—voltage standing-wave ratio
VTVM—vacuum-tube voltmeter
VUAC—VHF/UHF Advisory Committee
VUCC—VHF/UHF Century Club
VXO—variable-frequency crystal oscillator

W

W—watt (kg m^{2}s^{-3}), unit of power
WAC—Worked All Continents
WAS—Worked All States
WBFM—wide-band frequency modulation
WEFAX—weather facsimile
Wh—watthour
WPM—words per minute
WRC—World Radio Conference
WVDC—working voltage, direct current

X

X—reactance
XCVR—transceiver
XFMR—transformer
XIT—transmitter incremental tuning
XO—crystal oscillator
XTAL—crystal
XVTR—transverter

Y

Y—crystal; admittance
YIG—yttrium iron garnet

Z

Z—impedance; also see UTC

5BDXCC—Five-Band DXCC
5BWAC—Five-Band WAC
5BWAS—Five-Band WAS
6BWAC—Six-Band WAC

°—degree (plane angle)
°C—degree Celsius (temperature)
°F—degree Fahrenheit (temperature)
α—(alpha) angles; coefficients, attenuation constant, absorption factor, area, common-base forward current-transfer ratio of a bipolar transistor
β—(beta) angles; coefficients, phase constant current gain of common-emitter transistor amplifiers
γ—(gamma) specific gravity, angles, electrical conductivity, propagation constant
Γ—(gamma) complex propagation constant
δ—(delta) increment or decrement; density; angles
Δ—(delta) increment or decrement determinant, permittivity
ε—(epsilon) dielectric constant; permittivity; electric intensity
ζ—(zeta) coordinates; coefficients
η—(eta) intrinsic impedance; efficiency; surface charge density; hysteresis; coordinate
θ—(theta) angular phase displacement; time constant; reluctance; angles
ι—(iota) unit vector
κ—(kappa) susceptibility; coupling coefficient
λ—(lambda) wavelength; attenuation constant
Λ—(lambda) permeance
μ—(mu) permeability; amplification factor; micro (prefix for 10^{-6})
μC—microcomputer
μF—microfarad
μH—microhenry
μP—microprocessor
ξ—(xi) coordinates
π—(pi) 3.14159
ρ—(rho) resistivity; volume charge density; coordinates; reflection coefficient
σ—(sigma) surface charge density; complex propagation constant; electrical conductivity; leakage coefficient; deviation
Σ—(sigma) summation
τ—(tau) time constant; volume resistivity; time-phase displacement; transmission factor; density
φ—(phi) magnetic flux; angles
Φ—(phi) summation
χ—(chi) electric susceptibility; angles
ψ—(psi) dielectric flux; phase difference; coordinates; angles
ω—(omega) angular velocity 2pf
Ω—(omega) resistance in ohms; solid angle

Allocation of International Call Signs

Call Sign Series	Allocated to	Call Sign Series	Allocated to	Call Sign Series	Allocated to
AAA-ALZ	United States of America	† ERA-ERZ	Moldova	J6A-J6Z	Saint Lucia
AMA-AOZ	Spain	† ESA-ESZ	Estonia	J7A-J7Z	Dominica
APA-ASZ	Pakistan	ETA-ETZ	Ethiopia	† J8A-J8Z	St. Vincent and the Grenadines
ATA-AWZ	India	EUA-EWZ	Belarus		
AXA-AXZ	Australia	† EXA-EXZ	Kyrgyzstan	KAA-KZZ	United States of America
AYA-AZZ	Argentina	† EYA-EYZ	Tajikistan		
A2A-A2Z	Botswana	† EZA-EZZ	Turkmenistan	LAA-LNZ	Norway
A3A-A3Z	Tonga	† E2A-E2Z	Thailand	LOA-LWZ	Argentina
A4A-A4Z	Oman	FAA-FZZ	France	LXA-LXZ	Luxembourg
A5A-A5Z	Bhutan	GAA-GZZ	United Kingdom of Great Britain and Northern Ireland	† LYA-LYZ	Lithuania
A6A-A6Z	United Arab Emirates			LZA-LZZ	Bulgaria
				L2A-L9Z	Argentina
A7A-A7Z	Qatar	HAA-HAZ	Hungary	MAA-MZZ	United Kingdom of Great Britain and Northern Ireland
A8A-A8Z	Liberia	HBA-HBZ	Switzerland		
A9A-A9Z	Bahrain	HCA-HDZ	Ecuador		
BAA-BZZ	China	HEA-HEZ	Switzerland	NAA-NZZ	United States of America
CAA-CEZ	Chile	HFA-HFZ	Poland		
CFA-CKZ	Canada	HGA-HGZ	Hungary	OAA-OCZ	Peru
CLA-CMZ	Cuba	HHA-HHZ	Haiti	ODA-ODZ	Lebanon
CNA-CNZ	Morocco	HIA-HIZ	Dominican Republic	OEA-OEZ	Austria
COA-COZ	Cuba			OFA-OJZ	Finland
CPA-CPZ	Bolivia	HJA-HKZ	Colombia	OKA-OLZ	Czech Republic
CQA-CUZ	Portugal	HLA-HLZ	South Korea	OMA-OMZ	Slovak Republic
CVA-CXZ	Uruguay	HMA-HMZ	North Korea	ONA-OTZ	Belgium
CYA-CZZ	Canada	HNA-HNZ	Iraq	OUA-OZZ	Denmark
C2A-C2Z	Nauru	HOA-HPZ	Panama	PAA-PIZ	Netherlands
C3A-C3Z	Andorra	HQA-HRZ	Honduras	PJA-PJZ	Netherlands Antilles
C4A-C4Z	Cyprus	HSA-HSZ	Thailand		
C5A-C5Z	Gambia	HTA-HTZ	Nicaragua	PKA-POZ	Indonesia
C6A-C6Z	Bahamas	HUA-HUZ	El Salvador	PPA-PYZ	Brazil
* C7A-C7Z	World Meteorologica Organization	HVA-HVZ	Vatican City	PZA-PZZ	Suriname
		HWA-HYZ	France	P2A-P2Z	Papua New Guinea
		HZA-HZZ	Saudi Arabia	P3A-P3Z	Cyprus
C8A-C9Z	Mozambique	H2A-H2Z	Cyprus	† P4A-P4Z	Aruba
DAA-DRZ	Germany	H3A-H3Z	Panama	P5A-P9Z	North Korea
DSA-DTZ	South Korea	H4A-H4Z	Solomon Islands	RAA-RZZ	Russian Federation
DUA-DZZ	Philippines	H6A-H7Z	Nicaragua		
D2A-D3Z	Angola	H8A-H9Z	Panama	SAA-SMZ	Sweden
D4A-D4Z	Cape Verde	IAA-IZZ	Italy	SNA-SRZ	Poland
D5A-D5Z	Liberia	JAA-JSZ	Japan	• SSA-SSM	Egypt
D6A-D6Z	Comoros	JTA-JVZ	Mongolia	• SSN-SSZ	Sudan
D7A-D9Z	South Korea	JWA-JXZ	Norway	STA-STZ	Sudan
EAA-EHZ	Spain	JYA-JYZ	Jordan	SUA-SUZ	Egypt
EIA-EJZ	Ireland	JZA-JZZ	Indonesia	SVA-SZZ	Greece
† EKA-EKZ	Armenia	J2A-J2Z	Djibouti	S2A-S3Z	Bangladesh
ELA-ELZ	Liberia	J3A-J3Z	Grenada	† S5A-S5Z	Slovenia
† EMA-EOZ	Ukraine	J4A-J4Z	Greece	S6A-S6Z	Singapore
EPA-EQZ	Iran	J5A-J5Z	Guinea-Bissau		

Call Sign Series	Allocated to	Call Sign Series	Allocated to	Call Sign Series	Allocated to
S7A-S7Z	Seychelles	† V4A-V4Z	Saint Kitts and Nevis	† Z2A-Z2Z	Zimbabwe
S9A-S9Z	Sao Tome and Principe	† V5A-V5Z	Namibia	† Z3A-Z3Z	Macedonia (Former Yugoslav Republic)
TAA-TCZ	Turkey	† V6A-V6Z	Micronesia		
TDA-TDZ	Guatemala	† V7A-V7Z	Marshall Islands	2AA-2ZZ	United Kingdom of Great Britain and Northern Ireland
TEA-TEZ	Costa Rica	† V8A-V8Z	Brunei		
TFA-TFZ	Iceland	WAA-WZZ	United States of America		
TGA-TGZ	Guatemala	XAA-XIZ	Mexico		
THA-THZ	France	XJA-XOZ	Canada	3AA-3AZ	Monaco
TIA-TIZ	Costa Rica	XPA-XPZ	Denmark	3BA-3BZ	Mauritius
TJA-TJZ	Cameroon	XQA-XRZ	Chile	3CA-3CZ	Equatorial Guinea
TKA-TKZ	France	XSA-XSZ	China		
TLA-TLZ	Central Africa	XTA-XTZ	Burkina Faso	• 3DA-3DM	Swaziland
TMA-TMZ	France	XUA-XUZ	Cambodia	• 3DN-3DZ	Fiji
TNA-TNZ	Congo	XVA-XVZ	Viet Nam	3EA-3FZ	Panama
TOA-TQZ	France	XWA-XWZ	Laos	3GA-3GZ	Chile
TRA-TRZ	Gabon	XXA-XXZ	Portugal	3HA-3UZ	China
TSA-TSZ	Tunisia	XYA-XZZ	Myanmar	3VA-3VZ	Tunisia
TTA-TTZ	Chad	YAA-YAZ	Afghanistan	3WA-3WZ	Viet Nam
TUA-TUZ	Ivory Coast	YBA-YHZ	Indonesia	3XA-3XZ	Guinea
TVA-TXZ	France	YIA-YIZ	Iraq	3YA-3YZ	Norway
TYA-TYZ	Benin	YJA-YJZ	Vanuatu	3ZA-3ZZ	Poland
TZA-TZZ	Mali	YKA-YKZ	Syria	4AA-4CZ	Mexico
T2A-T2Z	Tuvalu	† YLA-YLZ	Latvia	4DA-4IZ	Philippines
T3A-T3Z	Kiribati	YMA-YMZ	Turkey	† 4JA-4KZ	Azerbaijan
T4A-T4Z	Cuba	YNA-YNZ	Nicaragua	† 4LA-4LZ	Georgia
T5A-T5Z	Somalia	YOA-YRZ	Romania	4MA-4MZ	Venezuela
T6A-T6Z	Afghanistan	YSA-YSZ	El Salvador	4NA-4OZ	Yugoslavia
† T7A-T7Z	San Marino	YTA-YUZ	Yugoslavia	4PA-4SZ	Sri Lanka
† T9A-T9Z	Bosnia and Herzegovina	YVA-YYZ	Venezuela	4TA-4TZ	Peru
UAA-UIZ	Russian Federation	YZA-YZZ	Yugoslavia	* 4UA-4UZ	United Nations
		Y2A-Y9Z	Germany	4VA-4VZ	Haiti
† UJA-UMZ	Uzbekistan	ZAA-ZAZ	Albania	4XA-4XZ	Israel
† UNA-UQZ	Kazakhstan	ZBA-ZJZ	United Kingdom of Great Britain and Northern Ireland	* 4YA-4YZ	International Civil Aviation Organization
URA-UTZ	Ukraine				
† UUA-UZZ	Ukraine			4ZA-4ZZ	Israel
VAA-VGZ	Canada	ZKA-ZMZ	New Zealand	5AA-5AZ	Libya
VHA-VNZ	Australia	ZNA-ZOZ	United Kingdom of Great Britain and Northern Ireland	5BA-5BZ	Cyprus
VOA-VOZ	Canada			5CA-5GZ	Morocco
VPA-VSZ	United Kingdom of Great Britain and Northern Ireland			5HA-5IZ	Tanzania
				5JA-5KZ	Colombia
		ZPA-ZPZ	Paraguay	5LA-5MZ	Liberia
VTA-VWZ	India	ZQA-ZQZ	United Kingdom of Great Britain and Northern Ireland	5NA-5OZ	Nigeria
VXA-VYZ	Canada			5PA-5QZ	Denmark
VZA-VZZ	Australia			5RA-5SZ	Madagascar
† V2A-V2Z	Antigua and Barbuda			5TA-5TZ	Mauritania
		ZRA-ZUZ	South Africa	5UA-5UZ	Niger
† V3A-V3Z	Belize	ZVA-ZZZ	Brazil	5VA-5VZ	Togo

Call Sign Series	Allocated to	Call Sign Series	Allocated to	Call Sign Series	Allocated to
5WA-5WZ	Western Samoa	7QA-7QZ	Malawi	9HA-9HZ	Malta
5XA-5XZ	Uganda	7RA-7RZ	Algeria	9IA-9JZ	Zambia
5YA-5ZZ	Kenya	7SA-7SZ	Sweden	9KA-9KZ	Kuwait
6AA-6BZ	Egypt	7TA-7YZ	Algeria	9LA-9LZ	Sierra Leone
6CA-6CZ	Syria	7ZA-7ZZ	Saudi Arabia	9MA-9MZ	Malaysia
6DA-6JZ	Mexico	8AA-8IZ	Indonesia	9NA-9NZ	Nepal
6KA-6NZ	South Korea	8JA-8NZ	Japan	9OA-9TZ	Zaire
6OA-6OZ	Somalia	8OA-8OZ	Botswana	9UA-9UZ	Burundi
6PA-6SZ	Pakistan	8PA-8PZ	Barbados	9VA-9VZ	Singapore
6TA-6UZ	Sudan	8QA-8QZ	Maldives	9WA-9WZ	Malaysia
6VA-6WZ	Senegal	8RA-8RZ	Guyana	9XA-9XZ	Rwanda
6XA-6XZ	Madagascar	8SA-8SZ	Sweden	9YZ-9ZZ	Trinidad and Tobago
6YA-6YZ	Jamaica	8TA-8YZ	India		
6ZA-6ZZ	Liberia	8ZA-8ZZ	Saudi Arabia		
7AA-7IZ	Indonesia	† 9AA-9AZ	Croatia		
7JA-7NZ	Japan	9BA-9DZ	Iran		
7OA-7OZ	Yemen	9EA-9FZ	Ethiopia		
7PA-7PZ	Lesotho	9GA-9GZ	Ghana		

- • Half series
- * Series allocated to an international organization
- † Provisional allocation in accordance with RR2088

Index

FEEDBACK

Please use this form to give us your comments on this book and what you'd like to see in future editions.

Where did you purchase this book?
   ☐ From ARRL directly   ☐ From an ARRL dealer

Is there a dealer who carries ARRL publications within:
   ☐ 5 miles   ☐ 15 miles   ☐ 30 miles of your location?   ☐ Not sure.

License class:
   ☐ Novice   ☐ Technician   ☐ Technician Plus   ☐ General   ☐ Advanced   ☐ Extra

Name _____

Daytime Phone ( ) _____

Address _____

City, State/Province, ZIP/Postal Code _____

If licensed, how long? _____

Other hobbies _____

_____

Occupation _____

ARRL member? ☐ Yes ☐ No

Call Sign _____

Age _____

For ARRL use only		HRME
Edition	1 2 3 4 5 6 7 8 9 10 11 12	
Printing	1 2 3 4 5 6 7 8 9 10 11 12	

From _____

_____

_____

EDITOR, HAM RADIO MADE EASY
AMERICAN RADIO RELAY LEAGUE
225 MAIN STREET
NEWINGTON CT 06111-1494

···················· please fold and tape ····················

HAM RADIO
MADE EASY

PROOF OF
PURCHASE